Challenges in MACHINE GENERATION of Analytic Products from MULTI-SOURCE DATA

PROCEEDINGS OF A WORKSHOP

Linda Casola, *Rapporteur*

Intelligence Community Studies Board

Division on Engineering and Physical Sciences

The National Academies of
SCIENCES · ENGINEERING · MEDICINE

THE NATIONAL ACADEMIES PRESS
Washington, DC
www.nap.edu

THE NATIONAL ACADEMIES PRESS **500 Fifth Street, NW** **Washington, DC 20001**

This activity was supported by Contract No. 2014-14041100003-012 between the National Academy of Sciences and the Office of the Director of National Intelligence. Any opinions, findings, conclusions, or recommendations expressed in this publication do not necessarily reflect the views of any organization or agency that provided support for the project.

International Standard Book Number-13: 978-0-309-46573-1
International Standard Book Number-10: 0-309-46573-7
Digital Object Identifier: https://doi.org/10.17226/24900

Additional copies of this publication are available for sale from the National Academies Press, 500 Fifth Street, NW, Keck 360, Washington, DC 20001; (800) 624-6242 or (202) 334-3313; http://www.nap.edu.

Printed in the United States of America

Suggested citation: National Academies of Sciences, Engineering, and Medicine. 2017. *Challenges in Machine Generation of Analytic Products from Multi-Source Data: Proceedings of A Workshop*. Washington, DC: The National Academies Press. doi: https://doi.org/10.17226/24900.

PLANNING COMMITTEE ON THE INTELLIGENCE MACHINE ANALYTICS WORKSHOP

RAMA CHELLAPPA, University of Maryland, College Park, *Chair*
THOMAS DIETTERICH, Oregon State University
ANTHONY HOOGS, Kitware, Inc.
JOHN E. KELLY III, NAE,[1] International Business Machines Corporation
KATHLEEN McKEOWN, Columbia University
JOSEPH L. MUNDY, Vision Systems, Inc.

Staff

GEORGE COYLE, Senior Program Officer, *Study Director*
CHRIS JONES, Financial Officer
MARGUERITE SCHNEIDER, Administrative Coordinator
DIONNA ALI, Research Assistant
ADRIANNA HARGROVE, Senior Program Assistant/Financial Assistant

[1] Member, National Academy of Engineering.

INTELLIGENCE COMMUNITY STUDIES BOARD

DONALD M. KERR, Independent Consultant, *Chair*
JULIE BRILL, Microsoft Corporation
FREDERICK CHANG, NAE,[1] Southern Methodist University
TOMÁS DÍAZ DE LA RUBIA, Purdue University Discovery Park
ROBERT C. DYNES, NAS,[2] University of California, San Diego
ROBERT FEIN, McLean Hospital/Harvard Medical School
MIRIAM JOHN, Independent Consultant
ANITA JONES, NAE, University of Virginia
ROBERT H. LATIFF, R. Latiff Associates
MARK LOWENTHAL, Johns Hopkins University
MICHAEL MARLETTA, NAS/NAM,[3] University of California, Berkeley
L. ROGER MASON, JR., Noblis
ELIZABETH RINDSKOPF PARKER, State Bar of California
WILLIAM H. PRESS, NAS, University of Texas, Austin
DAVID A. RELMAN, NAM, Stanford University

Staff

ALAN SHAW, Director
ANDREW KREEGER, Program Officer
CHRIS JONES, Financial Officer
MARGUERITE SCHNEIDER, Administrative Coordinator
DIONNA ALI, Research Assistant
STEVEN DARBES, Research Assistant
ADRIANNA HARGROVE, Senior Program Assistant/Financial Assistant

[1] Member, National Academy of Engineering.
[2] Member, National Academy of Sciences.
[3] Member, National Academy of Medicine.

Acknowledgment of Reviewers

This Proceedings of a Workshop was reviewed in draft form by individuals chosen for their diverse perspectives and technical expertise. The purpose of this independent review is to provide candid and critical comments that will assist the National Academies of Sciences, Engineering, and Medicine in making each published proceedings as sound as possible and to ensure that it meets the institutional standards for quality, objectivity, evidence, and responsiveness to the charge. The review comments and draft manuscript remain confidential to protect the integrity of the process.

We thank the following individuals for their review of this proceedings:

Anthony Hoogs, Kitware, Inc.,
Kathleen McKeown, Columbia University,
John Montgomery, NAE,[1] Naval Research Laboratory (retired),
Noah A. Smith, University of Washington, and
Peter Weinberger, Google, Inc.

Although the reviewers listed above provided many constructive comments and suggestions, they were not asked to endorse the content of the proceedings nor did they see the final draft before its release. The review of this proceedings was overseen by Edward W. Felten, NAE, Princeton University, who was responsible for making certain that an independent examination of this proceedings was carried out in accordance with standards of the National Academies and that all review comments were carefully considered. Responsibility for the final content rests entirely with the rapporteur and the National Academies.

[1]Member, National Academy of Engineering.

Contents

1

Introduction

The Intelligence Community Studies Board (ICSB) of the National Academies of Sciences, Engineering, and Medicine convened a workshop on August 9-10, 2017, to examine challenges in machine generation of analytic products[1] from multi-source data. With funding from the Office of the Director of National Intelligence (ODNI), the ICSB established a planning committee (see p. v and Appendix A) to develop the workshop agenda (see Appendix B). The workshop statement of task is shown in Appendix C. More than 300 people registered to participate in the workshop either in person or online.

Workshop speakers and participants discussed research challenges related to machine-based methods for generating analytic products and for automating the evaluation of these products, with special attention to learning from small data,[2] using multi-source data, adversarial learning,[3] and understanding the human–machine relationship. During the presentations and discussion sessions, attendees were asked to address the following questions, with particular emphasis on their role for the intelligence community:

- What are the primary issues?
- What is the research knowledge base?
- What are the current and "next level" of key performance metrics?
- What is the "level after next" of expected research and development performance?
- What are the requisite enabling technologies?
- How can the government best prepare the scientific workforce to enhance discovery in this area?
- What are the technical objectives and metrics needed for success?

---

[1] For the purposes of this report, ODNI defines "machine generation of analytic products" as the use of computer algorithms to quickly and accurately discover, interpret, and relay actionable information or insights to a consumer.

[2] Small data are data in low volumes, structured varieties, and batch velocities (See IBM Big Data and Analytics Hub, Infographics & Animations, "Taming Big Data: Small Data vs. Big Data," http://www.ibmbigdatahub.com/infographic/taming-big-data-small-data-vs-big-data, accessed August 9, 2017).

[3] Adversarial learning refers to the intersection of computer security and machine learning ((See IBM Big Data and Analytics Hub, Infographics & Animations, "Taming Big Data: Small Data vs. Big Data," http://www.ibmbigdatahub.com/infographic/taming-big-data-small-data-vs-big-data, accessed August 9, 2017).

This proceedings is a factual summary of what occurred at the workshop. The planning committee's role was limited to organizing and convening the workshop. The views contained in this proceedings are those of the individual workshop participants and do not necessarily represent the views of the participants as a whole, the planning committee, or the National Academies of Sciences, Engineering, and Medicine. In addition to the summary provided here, materials related to the workshop, including videos of presentation and discussion sessions, can be found at http://sites.nationalacademies.org/DEPS/icsb/DEPS_180349 or https://vimeo.com/album/4730093.

2

Session 1: Plenary

OPENING REMARKS

David M. Isaacson, Office of the Director of National Intelligence
Rama Chellappa, University of Maryland, College Park
George Coyle, National Academies of Sciences, Engineering, and Medicine

David Isaacson, Office of the Director of National Intelligence (ODNI), opened the workshop by providing an overview of ODNI as well as its expectations for the workshop. He explained that ODNI is developing a framework to identify emerging intelligence challenges, which could help generate investment guidance for research and development. The general information pipeline in the intelligence community includes tasking, collection, processing, exploitation, and product dissemination stages. Each stage is labor-intensive and time-consuming, with human-performed tasks introducing latency into the pipeline. New machine analysis techniques have the potential to improve decision making and quicken product deployment in the intelligence community by automating processes and reducing the number of bottlenecks that currently exist in the pipeline.

Noting ODNI's preliminary work in this area, Isaacson discussed the Xpress Challenge,[1] a joint competition sponsored by ODNI and the Office of the Under Secretary of Defense for Intelligence (OUSD(I)). Three hundred eighty-seven people from 42 countries registered for the challenge, and 15 of these registrants submitted solutions that are currently under review by ODNI. The task was to create an algorithm for an analytic product, using established intelligence community formatting and evaluation criteria, that could address an intelligence question and be used to aid policy makers and war fighters. Upon completion of the Xpress Challenge, ODNI and OUSD(I) created the Xtend Challenge,[2] scheduled to open in Fall 2017. Xtend Challenge participants will be asked to develop methods for machine evaluation of analytic products, which could be used to improve the quality of traditional human-generated analytic products. Isaacson believes these challenges reveal information about the current state of the art and what is next (and "after next") in innovation for the intelligence community.

A central topic of exploration for this workshop, Isaacson explained, is how machines and human analysts can share analytical tasks to improve efficiency and reliability. It is also important, he noted, to ensure that fund-

[1] The website for the Xpress Challenge is https://www.innocentive.com/ar/challenge/9933982, accessed August 24, 2017.

[2] When open, the Xtend Challenge will be posted on the following Challenge.gov website at https://www.challenge.gov/agency/office-of-director-of-national-intelligence/.

ing for such research is adequate but unduplicated among government sponsors. In light of these issues, Isaacson requested that the workshop participants create a capability technology matrix (see Appendix D[3]) that would connect current and future research challenges with current or planned investments for relevant federal agencies. Isaacson encouraged speakers and participants to be forward-thinking in their discussions of promising research opportunities and effective technologies.

Workshop chair Rama Chellappa, University of Maryland, College Park, commented that certain issues are important for research and innovation in national security, such as human and/or machine bias in decision-making tasks, adversarial input, product quality, result accuracy, multi-source data, social media data sources, and the balance of physics and geometry with data.[4] George Coyle, National Academies, noted that each speaker would be asked to identify technologies (either hardware or software) enabling the artificial intelligence/machine learning capabilities addressed in their respective presentations, and federal government representatives would offer an overview of the state of artificial intelligence/machine learning investment in their respective agencies.

PROGRESS IN MACHINE LEARNING

Tom Dietterich, Oregon State University

Tom Dietterich, Oregon State University, provided a brief overview of progress in machine learning over the past 10–20 years. Supervised learning is the process of training a computer to identify or classify an unknown input based on some previously observed training examples. For example, supervised learning can train a handwriting recognition system to identify letters and digits based on previous handwriting samples that have been labeled accordingly.

Dietterich described two fundamental paradigms that have been explored in machine learning: the probabilistic modeling paradigm and the end-to-end function learning paradigm. The learning process used in the probabilistic modeling paradigm involves designing and fitting a probabilistic model to the data and performing probabilistic inference to make a prediction. For the end-to-end function learning paradigm, a space of parameterized functions is defined and then the best setting of the parameters is determined by solving an optimization problem.

During the past 15–20 years, Dietterich noted, there has been much progress in both paradigms. The probabilistic modeling space has expanded with the development of (1) rich probabilistic models such as complex mixture models, (2) non-parametric models that allow the number of parameters to increase or decrease based on the amount of available data, and, most recently, (3) probabilistic programming.[5] Dietterich explained that there have been corresponding innovations in probabilistic inference methods, including in belief propagation, variational inference, Markov chain Monte Carlo, and Hamiltonian Monte Carlo techniques. Probabilistic modeling can help represent knowledge related to the problem at hand, reason about latent variables (which is crucial for the intelligence community), and generate calibrated assessments of uncertainty in the conclusions.

Dietterich described similar successes in function learning over the past two decades. Complex and expressive function classes have been developed including support vector machines, decision tree ensembles, the multi-layer perceptron, convolutional neural networks, long- and short-term memory networks, and residual networks. Dietterich added that experimentation with various models continues today at a rapid pace. He noted that creating language to describe models is only a starting point; the next step is to solve the optimization problem of fitting

[3] A diligent attempt was made to develop a full Capability Technology Matrix. This proved too aggressive to accomplish within the scope of a 2-day workshop. Nevertheless, each expert speaker was asked to present their thoughts on near- and long-term enabling technologies they believed would be useful to intelligence community analysts. These technologies are presented as a Capability Technology Table in Appendix D. Also, not all speakers responded to the request to identify enabling technologies. This caveat applies to all further references to the matrix in this workshop proceedings.

[4] The balance of physics and geometry with data was discussed in greater detail in the presentations of Joseph Mundy and Rama Chellappa, both summarized in Chapter 3 of this proceedings.

[5] An example of a probabilistic programming language is Church, a Scheme-like language that allows a probability distribution to be described by writing a program. Stan is a probabilistic programming language that has been embraced by the statistics community. For more information about probabilistic programming languages, see http://probabilistic-programming.org/wiki/Home, accessed September 4, 2017.

these models. Progress in this area has been made in large part owing to the collaborations between machine learning experts and operations research and optimization theory experts. With sufficient data, Dietterich explained, function learning can be more accurate than probabilistic modeling, in part because fewer modeling assumptions are made and less effort is required for model development.

Despite the many strengths of the probabilistic modeling paradigm and the end-to-end function learning paradigm, Dietterich explained that there are weaknesses related to the assumption of stationarity; need for an abundance of data; data collection process that is frequently, but often unnecessarily, biased; brittleness of end-to-end training; and need for new methods for verification, validation, and monitoring. To address this last weakness, Dietterich suggested increased collaboration among the software engineering, reliability engineering, and machine learning communities.

Concluding his presentation, Dietterich listed the machine learning community's current capabilities to (1) apply deep learning to signals-type data if they are stationary and if there is enough training data; and (2) adapt quickly to new problems using fine tuning and, to some extent, anomaly detection. In the next 3 to 5 years, Dietterich expects to see further improvement in the following:

- Open category classification,
- Automatic detection of biased and untrusted data sources,
- Anomaly detection on time-varying and network data, and
- Initial methods for validation and system monitoring.

In the longer term, he hopes there will be advances in the following:

- Defense against adversarial examples,
- Integration of large knowledge bases with machine learning,
- Use of meta reasoning to develop and test hypotheses about data source reliability, and
- Multi-scale machine learning.

Devanand Shenoy, Department of Energy, asked how much learning is impacted by computing power. Dietterich responded that computing power is critical to processing more data. This has motivated companies to develop new specialized chips. For example, Google is developing tensor processing units. He added that hardware improvement is an area in which much progress can be made. An audience participant observed that Dietterich's presentation primarily focused on machine learning within a particular modality of data and wondered what that implies about the readiness of machine learning to handle multi-modal data. Dietterich agreed with the participant that machine learning develops point solutions to single data sources, so the machine learning community is not the best place to look for solutions related to multi-modal data. In response to a question from an audience participant about multi-scale methods, Dietterich noted that such methods are needed to detect changes at many scales and adjust accordingly.

INDUSTRY PERSPECTIVE

Josyula R. Rao, IBM Corporation

Josyula Rao, IBM Corporation, focused his presentation on how machine learning has been incorporated in the security sector as well as how it may be utilized in the future. He shared that in 2016, more than 4 billion records were breached across a variety of industries. He added that while a system can be breached within hours or minutes, it can take weeks or months to detect a breach and costs, on average, $7 million per breach in the United States. He explained that because each security tool addresses a different kind of problem, there are too many tools from which security officers can choose, thereby increasing the difficulty of providing visibility. Machine learning techniques can reduce this fragmentation and also address the skills gap that exists in the security workforce.

Rao provided an overview of a typical cybersecurity attack. Many attacks, he explained, combine a social engineering attack with an advanced persistent threat, resulting in the implantation of malware in the enterprise. The usual attack starts by spear phishing a vulnerable user; then, malware is downloaded and a back door is installed on a vulnerable machine. Malware moves laterally within the enterprise, subverting more machines until it locates the target and exfiltrates the information. Rao remarked that a new approach toward security control that can address these many vulnerabilities is needed in order to make better cybersecurity decisions.

Today, security analysts are employed in security operation centers to monitor activities within enterprise networks, but these analysts have an increasingly difficult, time-consuming, and often mundane task. After it detects a threatening activity, the enterprise has to work backward and identify affected users and devices, which, Rao emphasized, takes too long. A process that takes hours instead of months to respond to an attack is needed, but industry is only just starting to explore the use of technologies with this capability.

Rao presented the new life cycle of security defense: detect an event, distinguish between anomalous and malicious, consult threat intelligence (using natural language processing techniques), connect external threat intelligence with the internal threat to begin reasoning, connect the business operations, and provide automation. Rao added that the security industry has used advanced visualizations (e.g., vision and speech) and adversarial machine learning techniques to try to improve their approaches. He described an emerging data-driven approach (based on machine learning) that complements knowledge-driven approaches; this "man–machine symbiosis" is what will move the security field forward, according to Rao. Rao highlighted the ways in which machine learning techniques can be used to classify unstructured data and build defenses using fine-grained parameters around those determined to be high-value assets. He noted, however, that there are both organizational challenges and big data challenges associated with this process (e.g., volume, velocity, variety, veracity, scalability, and accuracy). He added that industry would benefit from more attention to offense instead of dedicating all of its efforts toward defense against security breaches.

Rao provided a few examples of useful data-driven approaches and models for security, including using rule-based systems that study generic attack patterns, viewing domain name server traffic to identify evasive cohort behaviors of botnets, beaconing detection, looking at web access patterns, predicting malicious URLs, passively exploring networks, and identifying device anomaly detections. He explained that adversaries use "fast-fluxing" (i.e., changing an IP address) and domain name generation to evade detection.

For insider threats, Rao explained that user activity monitoring is helpful. Big data approaches can be used to identify at-risk users, discover risky behavior, and uncover fraudulent activity. He added that social media monitoring, in combination with cognitive analytics and data mining, is also useful for identifying and predicting abnormal behaviors in an organization.

Rao explained that because there is an abundance of unstructured data available, it is necessary to build a security ontology in order to better understand observations as they relate to the realities of the external world. He also noted that artificial intelligence techniques can be used to secure cognitive systems. Rao concluded by reiterating that the human will never be displaced from security measures; the best approach is to combine data-driven intelligence with knowledge-driven intelligence.

Shenoy asked what can be done with bitcoin technology to eliminate its potential vulnerability. Rao noted that one idea is to remove the anonymity associated with using bitcoin, but neither academia nor industry is likely to support such a controversial solution. Anthony Hoogs, Kitware, Inc., asked if there has been any achievement in which a new attack is anticipated *before* it happens. Rao explained that most attackers can circumvent an enterprise's security controls, but if an enterprise can get to attackers before they *access the data*, that would be considered a predictive success because the *damage* has been prevented. Kathleen (Kathy) McKeown, Columbia University, asked what kind of ground truth data can be used to learn in different and new security scenarios that arise. Rao noted that the action has shifted to profiling what is normal for a user and forming a loose boundary around those activities and participating communities. If there is an attack, it will traverse those different communities, so then the anomalous activity must be identified through causality tracking. An audience participant asked how Rao's solution recognizes something that seems abnormal as normal. Rao said that this situation is called "drift detection," and it is relatively easy to track (e.g., in the case of an employee who accesses private information because he has new privileges from a promotion, not because of an intent to breach security).

OPERATIONAL PERSPECTIVE: PROJECT MAVEN

Travis W. Axtell, Office of the Under Secretary of Defense for Intelligence

Travis Axtell, Office of the Under Secretary of Defense for Intelligence (OUSD(I)), opened his presentation by encouraging workshop participants to watch Lt. Gen. Jack Shanahan's keynote address on Project Maven from the 2017 Conference on Geospatial Intelligence.[6] Axtell noted that, in collaboration with international defense partners, the science and technology community in the U.S. Department of Defense (DOD) is already (1) using machine learning in both limited and full operation systems, (2) applying machine learning concepts to help analysts downselect and upselect pieces of information, (3) planning an artificial intelligence workshop, and (4) expanding the artificial intelligence/machine learning practitioner workforce beyond the research organizations. Axtell emphasized that workforce is as important as the algorithms and the methods themselves when discussing the future of machine learning.

Quoting guidance from former Deputy Secretary of Defense Robert O. Work, Axtell explained that DOD must "do much more and move much faster . . . to take advantage of recent and future advances in [artificial intelligence, big data, and deep learning]" and has thus "establish[ed] the Algorithmic Warfare Cross-Functional Team [Project Maven] to accelerate DOD's integration of big data and machine learning . . . to turn the enormous volume of data available to DOD into actionable intelligence and insights at speed."[7] He noted that Project Maven expands research in artificial intelligence and increases the speed at which warfighters can be supported, jobs can be created, and automation engines can be developed for difficult problems. Axtell added that Project Maven will also work with data needed for wide-area motion imagery processing. He introduced five critical areas associated with Project Maven's plan: (1) develop a data-labeling enterprise, (2) apply neural networks for computer vision, (3) enhance computing power, (4) establish program of record integration, and (5) increase user engagement.

Axtell next moved to a brief discussion of ontology. He acknowledged that ontologies[8] can be a reasonable starting point but explained that it can be an outmoded way of thinking that may not be useful for solving current problems. Instead, he suggested that researchers label the data based on what is in the data and allow the ontology to grow from there. He added that it is also necessary to establish a pipeline to the program of record so that updates are available often. This pipeline includes data pre-processing, data labeling, algorithm development, and algorithm integration.

Axtell highlighted the mission areas of the six central activities with Project Maven: (1) computer vision, (2) document media exploitation, (3) collection management, (4) all-source analysis, (5) targeting, and (6) indications and warning. He emphasized that the need to utilize available masses of structured, labeled data sets extends beyond the intelligence community, throughout the entire DoD, and beyond the National Security agencies.

Opening the discussion section, Joseph Mundy, Vision Systems, Inc., asked how to best transition technology (e.g., by providing toolkits and computing infrastructure). Axtell explained that specific methods to improve transition are still being evaluated but hopefully technology integration will be more customized to each program office and to each task in the future. Axtell also suggested that interesting business relationships may emerge as a result of Project Maven, as it is crucial to partner with a company that has a long-term relationship with and understanding of the government's processes when transitioning to new technologies.

Jay Hughes, Space and Naval Warfare Systems Command, asked if machine learning or deep learning algorithmic applications can address the time delay associated with the contract process. Axtell agreed that cycles of slow responses are barriers to progress. Insights about the latest state of the art shared at this workshop can be useful in improving DOD's acquisition processes, Axtell continued. McKeown added that there are funding opportunities in the legal field to explore how machine learning techniques can be better utilized to analyze the contract process.

---

[6] The address can be viewed at Vimeo, "Keynote: Lt. Gen. John N.T. "Jack" Shanahan, Director for Defense Intelligence, Warfighter Support, OUSD(I)," https://vimeo.com/220699218, accessed August 24, 2017.

[7] R.O. Work, 2017, "Memorandum for Establishment of the Algorithmic Warfare Cross-Functional Team," https://www.govexec.com/media/gbc/docs/pdfs_edit/establishment_of_the_awcft_project_maven.pdf.

[8] For further discussion on ontologies, see summaries of the presentations of Josyula Rao, Joseph Mundy, Rama Chellappa, Kathy McKeown, and Dragomir Radev.

A participant asked Axtell if DOD uses crowdsourcing for data labeling. Axtell responded that DOD is evaluating various labeling strategies, including crowdsourcing. Some members of the workforce are eager to label data because they know it will help the military; the engineering science workforce will likely perform data labeling on a part-time basis depending on the type of data sets needed and the urgency of those needs. Axtell indicated that whichever strategy is adopted, a cultural change will be involved. Rao inquired about DOD's approach to securing both the talent and the technologies needed to execute the goals of Project Maven. Axtell said that the primary focus of Project Maven is to acquire and build relationships with existing commercial capabilities. Ultimately, DOD's objective is to maintain a competitive advantage. A participant asked if Axtell's team plans to engage the larger research and government funding communities for guidance. Axtell emphasized that traditional relationships (e.g., between the Defense Advanced Research Projects Agency and academia) will continue while new relationships are being formed.

3

Session 2: Machine Learning from Image, Video, and Map Data

LEARNING FROM OVERHEAD IMAGERY

Joseph L. Mundy, Vision Systems, Inc.

Joseph Mundy, Vision Systems, Inc., explained that the extreme resolution of satellite imagery is beneficial for applications in intelligence and defense. Most importantly, satellite imagery can surveil denied territories without needing control of the airspace, which is necessary if using aerial video.

Mundy shared that while space-based imaging began with the Corona Project in 1960, commercial applications of space imaging did not begin until the 1990s. Today, DigitalGlobe[1] is the largest provider of imagery to the Department of Defense (DOD). Commercial satellites range in size and cost, from the smallest costing approximately $100,000 and the largest costing approximately $50 million, according to Mundy. The Dove, which is similar in size to a shoebox, has a resolution of only about 3 m but can capture nearly every point on the surface of the Earth; larger satellites such as the WorldView4 have a resolution of approximately 30 cm but can provide views of only approximately 4 million km^2 per day (see Figure 3.1).

Mundy described the three phases of satellite imagery analysis: (1) Viewing collected images, assembling important information, and reporting up the chain of command; (2) Monitoring military facilities over a period of days and reporting on activities; and (3) Using statistics to determine aspects such as urban growth, infrastructure, population change, or natural disaster over a period of weeks to years. While investments in this third phase have not yet been substantial, Mundy suggested that this is an area in which machine learning technologies can have the greatest impact.

To provide context for current work, Mundy gave a brief overview of the Research and Development for Image Understanding Systems (RADIUS) program (1993–1997), funded by the Defense Advanced Research Projects Agency (DARPA) and the Central Intelligence Agency, which detected events from imagery based on three-dimensional models created manually. In 2017, Vision Systems, Inc., collaborated with the Air Force Research Laboratory, using the same technology from the RADIUS program, to learn patterns of normalcy in regions of interest and to discover anomalous deviation from normalcy by studying event triggers. Mundy noted that although the technology from the 1990s still worked in 2017, today's tools are much improved.

Mundy also revisited the three-dimensional reconstruction of the Corona era, when stereo pairs had to be

[1] The website for DigitalGlobe is https://www.digitalglobe.com, accessed August 24, 2017.

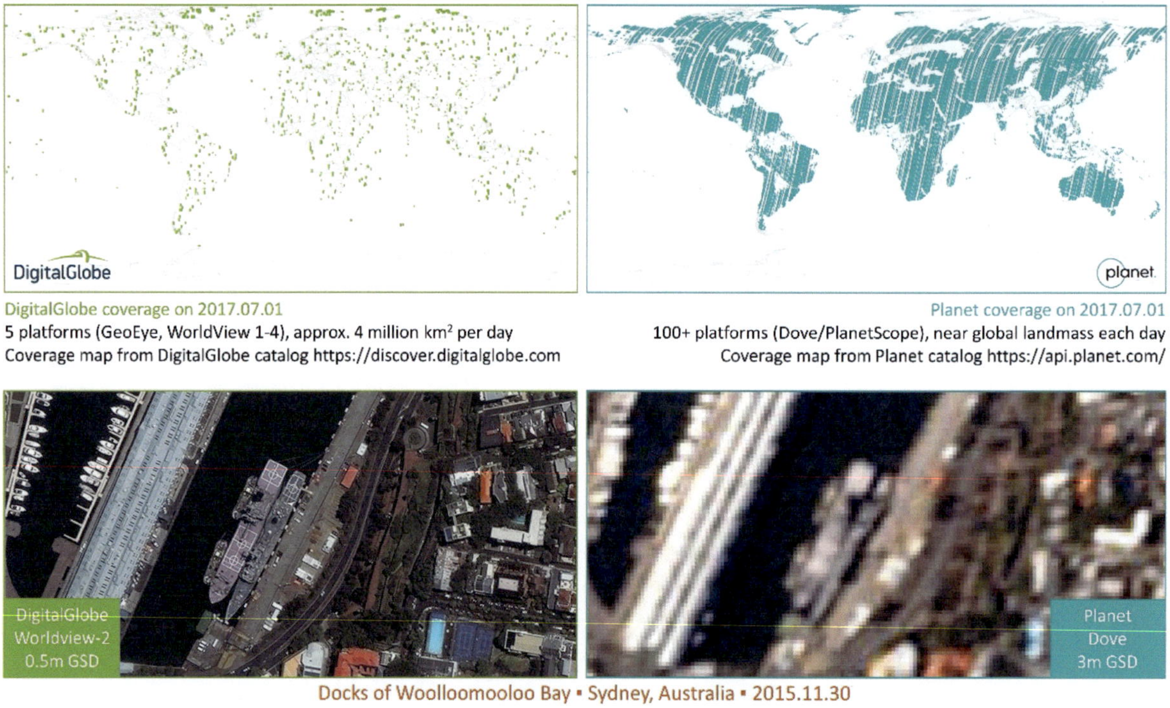

FIGURE 3.1 Coverage from the WorldView 1-4 (left) and 140+ Dove satellites, near global landmass each day (right). SOURCE: Joseph Mundy, presentation to the workshop, August 9, 2017. Images on the left courtesy of DigitalGlobe © 2017; images on the right courtesy of Planet Labs, Inc.

carefully constructed, and explained that, today, machine learning and classification can be used for multi-view reconstruction. He shared an application of this technology from Mosul, Iraq, in which two digital elevation models (made from images collected before and after the Fall of Mosul in 2014) were compared to detect structural changes. With the ability to construct three-dimensional data, researchers can develop a deeper understanding of the underlying geometry and create additional structural models of the data. Mundy also briefly described work on modeling three-dimensional buildings to emphasize the value of combining machine learning with more well-understood theoretical principles.

Mundy returned to a discussion of satellite imagery to highlight the issue of training data. One way to secure large quantities of training data is to generate examples of objects and object categories synthetically. If synthetically generated data are combined with manually labeled samples, a neural network can be trained to do the classification and to recognize the objects picked up in the satellite imagery. However, he pointed out that this is no simple task when there is an abundance of objects of interest to analyze. And even with the integration of adaptive learning, he continued, synthetic imagery still has its limitations. Mundy also highlighted Carnegie Mellon University's Terrapattern,[2] an open source tool for query-by-example in satellite imagery. For wide-area search, researchers use already existing crowdsourced information to identify areas of interest.

The future of satellite imagery, according to Mundy, lies in deep semantics: (1) Mature ontologies and automated extraction provide vast sources of linked knowledge; (2) Geographic labels are just the first generation of labeled training data; (3) Context from semantic web increases detection/classification performance; (4) Rapid satellite revisit and event ontologies will enable classification of four-dimensional concepts and processes; and

[2] The website for Terrapattern is http://www.terrapattern.com, accessed August 24, 2017.

(5) Automated discoveries expand linked knowledge to help "connect the dots." Mundy proposed that geometry will remain important for many years into the future because it serves as the basis for all architecture, and he suggested that the integration of theory-driven and data-driven approaches (by using machine learning algorithms to hypothesize and geometric reasoning to prove) is the best way forward.

Travis Axtell, Office of the Under Secretary of Defense for Intelligence, asked Mundy if he has seen any work done with occlusion in outdoor exterior scenes. Mundy explained that although he has not yet seen any work in that area, occlusion should be considered if the satellite imagery is limited. Axtell then asked Mundy if he has seen any work with multiple sources of intelligence related to having partially occluded three-dimensional point cloud scenes. Mundy said that he had not, but he highlighted some of Vision Systems's work in using three-dimensional scenes to fuse information from multiple sources. Rama Chellappa, University of Maryland, College Park, added that DARPA used multi-source integration for three-dimensional modeling in the late 1990s.

DEEP LEARNING FOR LEARNING FROM IMAGES AND VIDEOS: IS IT REAL?

Rama Chellappa, University of Maryland, College Park

Rama Chellappa, University of Maryland, College Park, agreed with Mundy that geometry and physics will always play important roles in the study of images and noted that ontology also remains useful because it provides a language for describing what is seen.

Chellappa explained that traditional computer vision approaches start with a problem, consider the best data (e.g., images, videos), identify the best features (e.g., interest points, stochastic models), and classify (e.g., using nearest neighbor rules, support vector machines). He emphasized that every problem has its own solution.

Chellappa explained that approaches for computer vision began to change with the increased use of neural networks in the late 1980s and early 1990s. Researchers discovered capabilities for face recognition, automatic target recognition, texture segmentation, optic flow computation, maxpooling, and image processing. Figure 3.2 shows the structure of LeNet, the first convolutional neural network. More recent developments in the field include graphical processing units, better understanding of non-linearities, and large annotated data sets. Chellappa noted that the deep convolutional neural networks (e.g., AlexNet) explored in 2012 are responsible for today's significant reduction in error rates (e.g., the top-1 and top-5 error rates have improved by 8 percent).

Deep learning has become more prevalent in computer vision, according to Chellappa. The Intelligence Advanced Research Projects Activity (IARPA) Janus Program[3] (2014–present) focuses on challenging problems related to verification, search, detection, and clustering. For example, the Janus team used different neural networks to try to optimize and extract deep features for face verification as well as three-dimensional models. He noted that some of Janus's data sets are available to the public.

Chellapa highlighted the "community of networks" that is formed when combining multiple networks that each use different training data. These more diverse learning experiences result in better generalizations. To Chellappa's surprise, deep learning has been working so well in Janus that the true accept rates have been higher than expected at 91 percent for 10-4 Federal Acquisition Regulation (FAR) for face verification.

Chellappa reiterated that computer vision uses reasoning to extend *beyond* pattern recognition. He emphasized that object detection, face detection, object recognition, and face recognition/verification have experienced improved performance and are essential for the computer vision community. Over the next 3–5 years, Chellappa hopes that advanced technologies will be able to address pose, illumination, and low-resolution challenges in unconstrained scenarios; handle full motion videos; enable sharing/transfer learning of features for different but somewhat related tasks; model/understand context and occlusion; handle a "reasonable" amount of noise in data; increase robustness to adversarial data; improve vision and language; and engage other multi-modal data.

Research prospects for the longer term include the following, according to Chellappa: theoretical investigation of why and when deep learning works; understanding of the minimum training set needed, given empirical distributions of the training set in the absence/presence of domain shift; advanced image formation, blur, geometry,

[3] The website for the Janus program is https://www.iarpa.gov/index.php/research-programs/janus, accessed August 24, 2017.

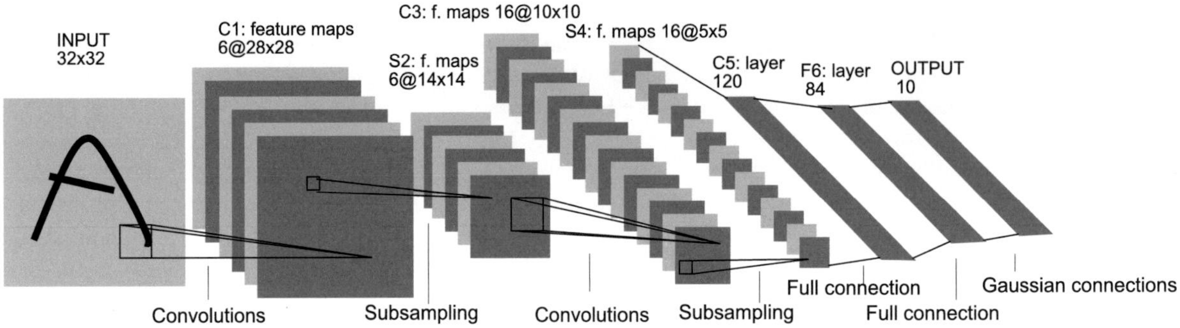

FIGURE 3.2 Structure of LeNet. SOURCE: © 2017 IEEE. Reprinted with permission from Y. LeCun, L. Bottou, Y. Bengio, and P. Haffner, Gradient-based learning applied to document recognition, *Proceedings of the IEEE*, 86(11):2278-2324, 1998.

shape, motion, texture, and occlusion models that can generalize using data; machine learning methods that generalize from small data and that deal with adversarial data; and domain knowledge and commonsense reasoning.

Chellappa added that students at the secondary and postsecondary levels are already doing machine learning and are eager to learn more. He expressed excitement about deep learning's potential now and its improvement in the future, especially given user-friendly software and promising data-labeling results.

Devanand Shenoy, Department of Energy, asked Chellappa if one can solve a problem using metrics, such as accuracy, instead of trying to solve it independent of the desired outcome. Chellappa responded that such metrics can be used with the training data, bearing in mind the generalizations to test data.

LEARNING ABOUT HUMAN ACTIVITIES FROM IMAGES AND VIDEOS

Anthony Hoogs, Kitware, Inc.

Anthony Hoogs, Kitware, Inc., noted that there is a hidden contribution in deep learning: the open source culture that it has fostered. Deep learning flourished in the computer vision community in 2014 with the development of the open source software framework Caffe.[4] Hoogs's Kitware, Inc., has been an open source company for nearly 20 years. Primarily funded by the U.S. government, Kitware now has 25 open source toolkits.

Hoogs explained that video domains vary greatly; for example, they include wide-area motion imagery, narrow field-of-view unmanned aerial vehicle video, ground-level surveillance, and internet/web videos. Most historic computer vision work in video has been in action recognition, where a video clip has a single action that is identified and labeled. Now, the focus often turns to action detection, which is useful when one does not know when and where in the video something is happening. This temporal search problem in surveillance is difficult, he stated, because, most of the time, nothing of interest is happening in the footage.

He noted that data sets drive research in vision because one has to build a data set that addresses the problem in need of a solution. In order to publish, researchers must have the best scores on an established data set or a new, interesting data set they have created. One downside of this paradigm, Hoogs said, is that if a training data set is too different from real-world data, published results will not translate into comparable performance on operational data.

Deep learning is playing a prominent role in video research, according to Hoogs. The application of deep learning to video explores both time and sequential data. Emerging work in this area is in the use of recurrent

[4] The website for Caffe is http://caffe.berkeleyvision.org, accessed August 31, 2017.

neural networks and long short-term memory networks.[5] There is also work in fusing parallel convolutional neural networks together. However, he noted that as most of the work is in supervised learning, there is relatively little work so far in unsupervised learning or generative modeling for video. There has also been little work on fusing video and audio and on model transfer. He explained that, overall, deep learning has had less impact so far on video analytics than on image recognition.

Hoogs provided an overview of various methods used to address action recognition challenges (e.g., cluttered backgrounds, occlusion, spatiotemporal uncertainty, and high inter-class variability). There are a number of data sets that help to address these challenges, but caution needs to be used to ensure the data match the problem at hand. For action detection (e.g., surveillance video), temporal localization and spatiotemporal localization methods can be used.

The DOD community has funded substantial work in analytics for Wide-Area Motion Imagery (WAMI), primarily before the deep learning revolution, according to Hoogs. However, complex activity detection and graph matching problems are areas in which deep learning has not yet shown much progress. He next introduced the IARPA Deep Intermodal Video Analytics (DIVA)[6] program for which Kitware creates the data set and develops an open source video analytics framework. DIVA works to improve activity detection in challenging conditions such as crowded scenes, low resolution, and low visibility. Hoogs mentioned the DARPA Video Image Retrieval and Analysis Tool (VIRAT) program[7] that addressed some of these issues pre-deep learning during the course of its surveillance research.

Hoogs remarked that commercial investment is substantial in unmanned aerial vehicle video research for small drones, particularly in domains such as precision agriculture, infrastructure inspection, and sports. He suggested that the intelligence community and DOD invest in problems that are unique and leverage current research as opposed to investing in research that industry is already conducting. For example, DOD and the intelligence community could focus on and invest in alternative sensing modalities such as infrared and hyperspectral since they are of less interest to commercial investors. Hoogs also encouraged the intelligence and defense communities to support open source frameworks and data sharing within relevant agencies and to engage academic researchers in problem solving. He encouraged the development of training methods for noisy, small training sets with heterogeneous labels, as well as focusing on coincident sensor fusion and fine-grained problems. Within 3 to 5 years, Hoogs expects substantial advances in video using deep learning, progress on the surveillance problem, and the integration of text and video for querying. Within 5 to 10 years, Hoogs hopes to see structure learning of deep networks on a large scale.

Sanjeev Agarwal, DOD Night Vision and Electronic Sensors Directorate, asked if a vision framework of live spatiotemporal video processing is needed to attack big vision problems such as human visual perception. Hoogs responded that, on the software side, people are building a video framework, but more systems work is needed. Chellappa remarked that the human perception view has to be integrated into this work. Hoogs noted that neuroscientists are already using computer vision research to map detailed functional brain studies. Jason Duncan, MITRE, noted that while deep learning shows great successes if there is good training data, it is still important to know how to solve problems with limited data sets.

---

[5] A long short-term memory network is a "specific recurrent neural network architecture that is well-suited to learn from experience to classify, process, and predict time series with time lags of unknown size" (See A. Azzouni and G. Pujolle, 2017, A long short-term memory recurrent neural network framework for network traffic matrix prediction, https://arxiv.org/abs/1705.05690).

[6] The website for the DIVA program is https://www.iarpa.gov/index.php/research-programs/diva, accessed August 29, 2017.

[7] For information about the VIRAT data set, see http://www.viratdata.org, accessed September 4, 2017.

4

Session 3: Machine Learning from Natural Languages

MACHINE LEARNING FROM TEXT: APPLICATIONS

Kathy McKeown, Columbia University

Kathy McKeown, Columbia University, explained that the natural language processing community uses machine learning techniques to extract information from text to solve problems. The framework for these machine learning techniques includes (1) the data, which is often labeled with the intended output; (2) the extraction of features from the text data using natural language techniques; and (3) the prediction of output. According to McKeown, researchers are interested in what type of data is available for learning, which features produce the most accurate predictions, and which machine learning techniques work best—supervised, unsupervised, and distant learning are currently used.

McKeown noted that core natural language processing technology enables part-of-speech tagging, sentence parsing, and language modeling (i.e., a statistical approach to identify word sequence probability). She detailed applications of natural language technology that use (1) interpretation (e.g., information extraction and sentiment analysis), (2) generation (e.g., summarization), and (3) analysis of social media (e.g., Tweets from gang members).

The purpose of information extraction, according to McKeown, is to extract concrete facts from text (i.e., named entities, relations, events) to create a structured knowledge base of facts. Event discovery and linking includes linking documents as well as linking dimension in the text to a real-world entity. The state of the art for the English language is relatively strong for named entities (89 percent accuracy), though there is room for improvement in both relations (59 percent accuracy) and events (63 percent accuracy). Commonly deployed machine learning methods include sequence labeling, neural nets, and distant learning, while features could be linguistic or based on similarity, popularity, gazeteers, ontologies, or verb triggers.

McKeown pointed out that since 2006, there has been substantial progress in entity discovery and linking as a result of the Defense Advanced Research Projects Agency (DARPA) Deep Exploration and Filtering of Text (DEFT) program.[1] Additional short-term goals for entity discovery and linking include improved name tagging, smarter collective inference, and resolution of true aliases and handles used as entity mentions. She added that

[1] For information about the DEFT program, see DARPA, "Deep Exploration and Filtering of Text (DEFT)," last updated November 13, 2015, https://opencatalog.darpa.mil/DEFT.html. See also the section on Boyan Onyshkevych's presentation in Chapter 5 of this proceedings.

more long-term goals would increase the number of languages (to 3,000) and entities (to 10,000), eliminate the use of training data, incorporate multimedia and streaming modes, and improve context awareness.

McKeown defined sentiment analysis as the ability to identify sentences that convey opinions, as well as the polarity of those opinions. Sentiment analysis allows for the detection of online influencers, agreements or disagreements in dialogues, and online conflicts, as well as for the understanding of argumentation. Methods for sentiment analysis are primarily supervised: (1) manually labeled or self-labeled data collections; (2) conditional random fields, support vector machines, and joint learning approaches; (3) recurrent neural networks; and (4) sentiment of words, word categories, parts of speech, and parse features. In the future, McKeown hopes that cross-lingual transfer of sentiment and sentiment toward targets will be possible.

Natural language processing is especially important when disasters happen around the world in places where researchers or aid workers neither know the language nor have access to machine translation, McKeown explained. The DARPA Low Resource Languages for Emergent Incidents (LORELEI) program[2] analyzes the subjective posts of everyday people and tries to identify sentiment and emotion in low resource languages. Cross-lingual sentiment models are used to leverage parallel and comparable corpora in low resource languages—multiple transfer model scenarios are created, target languages are engaged, and transfer models are created. These models, then, can be used to create sentiment analysis for languages with no labeled data. McKeown noted that the models perform well in English (80–85 percent) to detect polarity; however, when cross-lingual transfer is done, the success rate is much lower. For cross-lingual transfer, it is 40–50 percent; for sentiment toward targets, it is 59 percent.

By browsing and filtering news, journal articles, emails, and product reviews, McKeown noted that summarization enables quick determination of the relevance of these sources. Methods for summarization are typically extractive: sentences that are representative of the larger document are selected before unsupervised or supervised learning methods are applied. McKeown warns that the resulting summaries are not always coherent or accurate. If an abstractive method is used instead, pieces of the selected sentences are edited, compressed, or fused. There has been quite a bit of success with abstractive approaches, but they still need improvement. McKeown explained that the next step in summarization involves monitoring events over time using streaming data. For example, during disasters such as hurricanes or power outages, it is useful to take in relevant documents every hour and generate updates. Each disaster has a language model and geographic features. She added that summarization of personal experience from blogs or online narrative can also be used during a disaster and is a challenge for the next 5 years. Neural networks are a methodology that is just now being applied to summarization and holds promise for the future. The upcoming Intelligence Advanced Research Projects Activity (IARPA) Machine Translation for English Retrieval of Information in Any Language (MATERIAL) program will conduct search and summarization over low resource languages.

The third application McKeown discussed was a joint project between the social work and natural language processing communities to analyze tweets from gang members in Chicago.[3] This partnership was especially important because the qualitative analysis and annotation would not have been possible without the social workers' knowledge of language commonly used by gang members. A speech tagger and bilingual dictionary were created before the tweets could be classified and labeled as either "aggression" or "loss." Researchers observed that violence exacerbated after gang members taunted one another on social media. McKeown hopes that this case study will lead to increased community intervention to diffuse violence when aggression is detected. She noted that neural network models are needed to exploit large unlabeled data sets and wondered if these methods could be used to detect other instances of violence from nonstandard language.

McKeown's last topic of discussion was the identification of fake news through classification of statements and articles. She described a system that currently identifies fake news accurately 65 percent of the time, so there is more work to be done in this area as well. McKeown concluded with a discussion of technology capabilities over

---

[2] For more information about the LORELEI program, see B. Onyshkevych, "Low Resource Languages for Emergent Incidents (LORELEI)," DARPA, https://www.darpa.mil/program/low-resource-languages-for-emergent-incidents, accessed August 24, 2017. See also the section on Boyan Onyshkevych's presentation in Chapter 5 of this proceedings.

[3] T. Blevins, R. Kwiatkowski, J. Macbeth, K. McKeown, D. Patton, and O. Rambow, 2016, Automatically processing tweets from gang-involved youth: Toward detecting loss and aggression, pp. 2196-2206 in *Proceedings of COLING 2016, the 26th International Conference on Computational Linguistics: Technical*, December 11–17, available at http://aclweb.org/anthology/C/C16/.

the next 5 or more years. She explained that while the natural language processing community has strong capabilities for English and formal text, it needs to improve interpretation, analysis, and generation from multi-modal sources, informal texts, and streaming data. It is also important to be able to do learning without big data (e.g., in the case of low resource languages); better understand machine learning, discourse, context, and applications in multi-lingual environments without machine translation; fund research in robust and flexible language generation; and explore new application areas such as fake news, cyberbullying, and inappropriate content.

Jay Hughes, Space and Naval Warfare Systems Command, remarked that a common dictionary is needed to avoid miscommunication about concepts for which we use different words to describe. McKeown responded that the lack of a common vocabulary is of interest to the natural language processing community. It wants to capture cultural background, as well as perspective and ideology, to generate and understand language. A participant asked if fake news should be considered a computational linguistics problem, a graph analytics problem, or a natural language processing problem. McKeown said that it is a difficult problem, and a solution will likely require all technologies that can be brought to bear.

DEEP LEARNING FOR NATURAL LANGUAGE PROCESSING

Dragomir Radev, Yale University

Dragomir Radev, Yale University, noted that there has been a full-scale revolution in neural networks owing to the large amount of textual data (e.g., email, social media, news, spoken language) and the powerful hardware that are available. Radev remarked that many of the most fundamental natural language processing tasks are classification problems—for example, optical character recognition, spelling correction, part-of-speech tagging, word disambiguation, parsing, named entity segmentation and classification, sentiment analysis, and machine translation.

Radev described the traditional natural language processing pipeline. Words in a sentence are tagged with parts of speech and then parsed based on syntactic constituency or dependency (see Figure 4.1). To understand language, Radev explained, it is necessary to perform semantic analysis. After a sentence is converted into a logical form, world knowledge is used to perform inference and reach a conclusion about meaning. Current natural language processing methods include (1) vector semantics for dimensionality reduction or compositionality; (2) supervised learning using deep neural networks; and (3) learning architectures such as recurrent neural networks, long short-term memory networks, attention-based models, generative adversarial networks, reinforcement learning, and off-the-shelf libraries. Radev explained that, with vector semantics, words can be represented as vectors based on correspondence to nearby words, and meaning can be represented in vector form. He noted that this method also works when dealing with multilingual text. Compositional meaning can be expressed as a sum of vectors whereas non-compositional meaning (e.g., idioms) cannot.

Radev defined deep learning as neural networks with multiple layers and non-linearities. Depending on the number of layers in the neural network, deep learning can be used as a simple classifier or to learn more complex

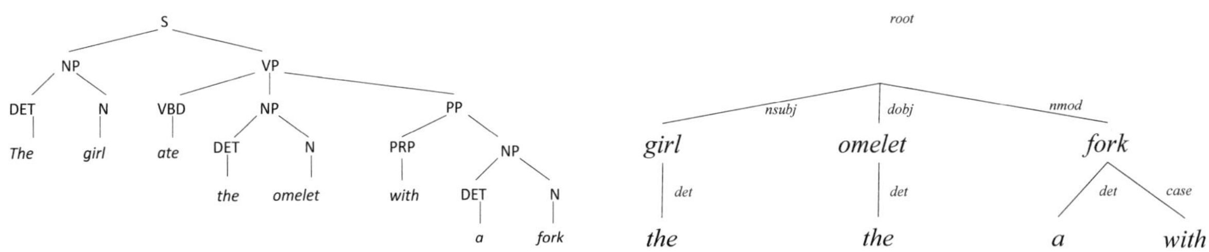

FIGURE 4.1 Examples of constituent parsing (left) and dependency parsing (right). SOURCE: Dragomir Radev, presentation to the workshop, August 9, 2017.

features. He added that deep learning works well with high-performance computers using graphical processing units and with large data sets. He shared that industry is investing billions in deep learning, in part because it has led to a significant improvement on many natural language processing tasks—for example, a 30 percent decrease in word error rate for speech recognition. Radev next highlighted multi-layer perceptrons, which can be used as classifiers. They are universal function approximators in the sense that they can compute any function that is consistent with the training data. The hidden layers between the input layer and the output layer are tasked with finding features of interest.

Radev gave a brief overview of some important work in the natural language processing and related communities:

- *AlexNet*. A neural network architecture used for computer vision.
- *Embeddings for words*. Used to represent singular and plural words, comparative and superlative adjectives, and male and female words, for example.
- *Autoencoder*. The simplest neural network that can perform dimensionality reduction—when the network is trained, the hidden dimensions find principle components for the lower dimensional representations of the data.
- *Embeddings for word senses*. Uses vectors to represent multiple meanings associated with individual words.
- *Recurrent neural networks*. Used to represent and make inferences about sequences of words, such as sentences.
- *Long short-term memory networks*. Addresses challenges with vanishing or exploding gradients in recurrent neural networks by controlling the flow of information through the network and enables long-distance dependency. Useful for language identification, paraphrase detection, speech recognition, handwriting, and music composition, for example.
- *Attention-based summarization*. Through the use of an attention mechanism, text can be processed and sentences simplified, enabling progress on tasks such as summarization.

Radev described the challenges that remain in natural language processing, including recognition of non-compositionality, incorporation of world knowledge, semantic parsing, discourse analysis, contextual comprehension, user modeling, database systems access, dialogue modeling, and text generation. There are also challenges with adversarial data when fake examples confuse the classifier.

Joseph Mundy, Vision Systems, Inc., asked why the attempts to introduce world knowledge in natural language processing have struggled to date. Radev noted that manually built ontologies of world knowledge are gradually fading, and the future is in automatically learned resources. Tom Dietterich, Oregon State University, commented that automatically structuring the architecture is important since deep networks require writing a program and that it is important to identify useful models and reusable program components for this new style of differentiable programming.

In conclusion, Radev encouraged the intelligence community to focus on the following research areas:

- Speech analysis,
- Massive multilinguality,
- One-shot learning,
- Understanding short texts,
- Interpretability of models,
- Deep learning personalization,
- Trust analytics, and
- Ethical language processing.

He also highlighted potential enabling technologies such as the following:

- Integrating deep learning and reinforcement learning,
- Automating architecture learning,

- Hyperparameter searching,
- Improving generative models,
- Using memory-based networks and neural Turing machines,
- Semi-supervised learning,
- Mixing and matching architectures, and
- Adopting privacy-preserving models.

MACHINE LEARNING FROM CONVERSATIONAL SPEECH

Amanda Stent, Bloomberg

Amanda Stent, Bloomberg, explained that conversation is a complex problem because while there are accepted rules that conversation participants follow, there are also myriad choices that can be made by these participants. Speech conversation researchers use advanced analytics and machine learning methods but are constrained by a lack of access to data appropriate for the complexity of the problem and thus sometimes work on artificial problems or with artificial data. As a result, according to Stent, speech and dialogue are *not* solved problems, and a divide exists between research and industrial application in this area.

She said that while conversation is often thought to occur between two or more humans, human–computer conversation is becoming more prevalent—the most common of which is slot-filling dialogue. Conversation is different from text or speech processing because (1) It is collaborative; (2) Its contributions are hearer-oriented, tentative, and grounded; and (3) It is incremental, multi-functional, and situated (i.e., there is always other context that must be taken into account). Stent noted that much research in conversation is focused on speech recognition uncertainty and errors because computer systems cannot yet handle these types of miscommunications.

Stent described three contexts in which machine learning methods would be useful to study conversation: exploration, understanding, and production. For instance, government agencies may be interested in the exploration component because it supports off-line and real-time data mining and analytics to identify issues with products or services. The machine learning methods that focus on understanding conversation support more advanced analytics and could be used to identify where a conversation veers off track, for example. Methods for production, on which researchers tend to focus most of their time, learn conversational agent models and task models. Across industry, these machine learning methods may be used for call center analytics, advertisement placement, customer service analytics, and problematic dialogue identification.

Stent added that there is a large gap between what researchers do and what industry does in the context of building conversational agents, which will hopefully begin to be addressed by the technology capabilities discussed in this workshop. She listed platforms that could serve as examples on which to build, including Microsoft Cognitive Services, Facebook, Slack, Twitter, Reddit, and various open source packages (e.g., rasa.ai and opendial). Evaluation of such systems is based on retrieval (for exploration), precision and recall (for understanding), and task completion and user satisfaction (for production). According to Stent, current threads of research in conversation include domain adaptation for dialogue state tracking and "end-to-end" models for chat dialogue, which predict the next utterance in a conversation, given the previous.

Stent emphasized that there are still many large problems in conversation. No general dialogue engine yet exists—an integrated model of comprehension and production at all levels of linguistic processing is needed. And researchers are still addressing basic architectural and modeling questions, which is very difficult to do with a limited amount of data. While a small number of large companies may have an abundance of data, these data may be of little use to researchers either because they are uninteresting or may contain personally identifiable information.

To advance innovation in conversation, Stent stated that sponsors are needed to support further research on machine learning for conversational analytics; this would give the conversation community access to more problems and more data. In the realm of producing conversational agents, Stent hopes to move beyond single-domain slot filling and context-free chat to the areas of situated, negotiation, and perpetual dialogue, for instance. The community could also expand slot-filling dialogue with more tasks, multiple domains, or multiple genres. Generally, Stent believes that methods combining advanced machine learning neural networks with programmatic control would be useful. She also encouraged the use of methods that include a "human-in-the-loop" so as to collect data and train the models more quickly. Further innovation could come from providing the community with shared portals for evaluation and crowdsourcing of data. Stent hopes that work on world knowledge induction and situational awareness will also be sponsored in the United States, as such work is already under way abroad. Stent would like to see researchers collaborate on issues of data availability and privacy.

Regarding the issue of data availability, a participant noted that some of the large companies that have access to an abundance of data may have more interesting data than we realize; they may just be inaccessible. Dietterich asked if planning technology is essential for the study of dialogue. Stent responded that there are many applications for which planning is crucial for the *task level* of the conversation. However, at the *conversation level*, Stent remains unconvinced that planning is necessary. Rama Chellappa, University of Maryland, College Park, asked if there will ever be a system that will convert a native language in real time so that everyone else in a country (e.g., India, which has 30 languages) can understand it. Stent responded that the Interspeech Sexennial Survey[4] estimated that such progress would be unlikely before the year 2050.

[4] Moore, R.K., and R. Marxer, 2016, Progress and prospects for spoken language technology: Results from four sexennial surveys, *Proceedings of INTERSPEECH*, September 8-12, https://pdfs.semanticscholar.org/99e8/7769b1d2b27f09012cdf31ea550d00be4c22.pdf.

5

Session 4: Learning from Multi-Source Data

SITUATIONAL AWARENESS FROM MULTIPLE UNSTRUCTURED SOURCES

Boyan Onyshkevych, Defense Advanced Research Projects Agency

Boyan Onyshkevych, Defense Advanced Research Projects Agency (DARPA), provided an overview of DARPA's mission, which is to maintain technological superiority of the U.S. military and prevent technological surprise from harming our national security. Thus, DARPA funds research in industry, academia, and the government laboratories on projects that will strengthen national security.

He introduced three DARPA programs that enable future analytics for both intelligence community and Department of Defense (DOD) applications (see Figure 5.1). In Deep Exploration and Filtering of Text (DEFT), Low Resource Languages for Emergent Incidents (LORELEI), and Active Interpretation of Disparate Alternatives (AIDA), there are media-specific transforms that look at input data (i.e., unstructured) in certain modalities (e.g., video, image, language) and produce a set of structured objects as output. In this case, the output that the end users work with is in the form of knowledge objects,[1] as opposed to language objects. Onyshkevych noted they are using these analytics to learn about the world as described in the corpus of available data by taking the knowledge objects and linking them to what is already known. This transition from document analytics to building a model about the world from the input data is where Onyshkevych suggested researchers dedicate their time moving forward. With such a model, researchers can conduct social network analysis, create entity relationship diagrams, and produce indications and warnings, without worrying about modality-specific problems.

Onyshkevych stated that the goal of the DEFT program, which is nearing conclusion, was to create automated capability to transform large volumes of unstructured text into structured data for use by human analysts and downstream analytics. DEFT's approach involves (1) finding and representing key information (e.g., entities, relations, events, sentiments, beliefs); (2) aggregating information from multiple sources, detecting inter-document relationships, and representing the information in a structured knowledge base; and (3) exploring and filtering entities, relations, and events from formal and informal inputs. He explained that in DEFT, the source material could include unstructured multilingual input in the form of questions. Document-level annotations are made

[1] "Knowledge object" refers to a "highly structured interrelated set of data, information, knowledge, and wisdom concerning some organizational, management, or leadership situation" (see Z.L. Berge and L.Y. Muilenburg, eds., 2013, *Handbook of Mobile Learning*, Routledge, New York).

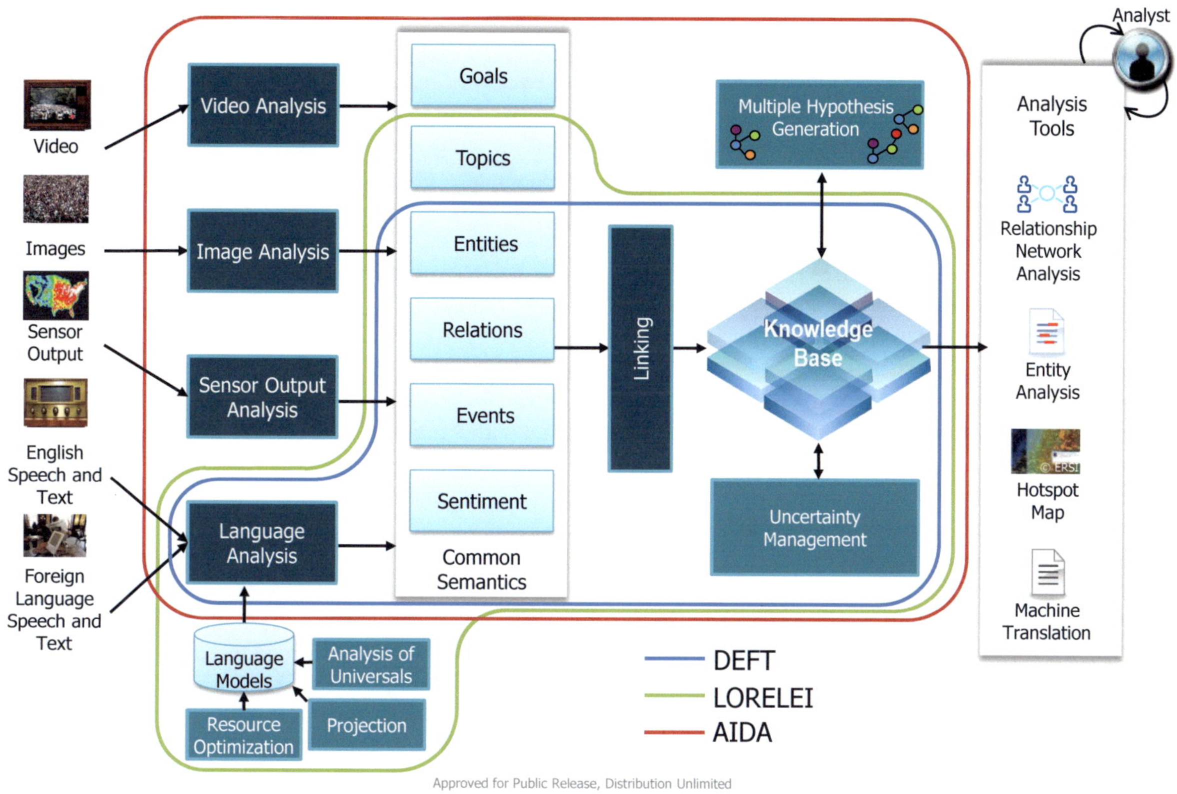

FIGURE 5.1 Three Defense Advanced Research Project Agency programs that enable future analytics for both intelligence community and Department of Defense applications. NOTE: AIDA, Active Interpretation of Disparate Alternatives; DEFT, Deep Exploration and Filtering of Text; LORELEI, Low Resource Languages for Emergent Incidents. SOURCE: Boyan Onyshkevych, presentation to the workshop, August 9, 2017.

and integrated with the core knowledge base assertions. Input is then linked, and inferences are made around the knowledge base through visualization or downstream analytics before machine-readable (i.e., structured) output emerges. And no matter the language of the source, the output should be the same. Over the past 5 years, DEFT–National Institute of Standards and Technology (NIST) open evaluation tasks have evolved from single document analysis to a composite picture of the world, allowing for joint inference and a single pipeline of information. Onyshkevych encouraged participants to consider attending DEFT's demonstration day in October, and he also noted that all DEFT code is open source.

The LORELEI Program, which is currently under way, uses the same model of converting unstructured input into structured, actionable output. However, this program specifically considers how to respond quickly to an emergent situation (e.g., humanitarian assistance and disaster relief missions) in a low resource language area. With more than 7,000 active languages in the world, and often no advance notice to secure interpreters or language technology resources, such work is crucial, Onyshkevych explained.

In the LORELEI scenarios, performers are given a language, must collect resources, and then are tested at checkpoints of 1 day, 1 week, and 1 month. There are three evaluation tasks: entity detection and linking, machine translation, and situation frames. The situation frames are designed to address the problem that during actual disaster relief, mission planners need to know what the need is, where it is located, how urgent it is, and whether it has been met upon completion of the mission. Onyshkevych said that one of the challenges lies in building this capability without supervised training data. The team relies on some distant supervision and language universals (i.e., languages are not randomly created, and many are related to one another in some way)—for example, only 2 of over 7,000 languages order a sentence with the object first, verb second, and subject last. Accordingly, the team built resource sets for 24 languages that are representative of many language families. The process the team uses is not fully automated; two humans (a native speaker of a language other than English, but with some English skills, and a linguistics expert) provide information into the system when learning about a language. However, this human participation is limited to only a few hours of hands-on time.

A third DARPA program, AIDA,[2] has not yet begun, but it plans to expand what DEFT did and what LORELEI is doing with unstructured and structured data. Onyshkevych commented that while LORELEI considers spoken and written language, AIDA will look at those plus image and video. From each situation, one can extract entities, events, and relations, link them to each other and to other knowledge sources, and then try to put together multiple hypotheses to explain as much of the situation as possible. Errors in processing, reporting, and misinterpretation are difficult to avoid, Onyshkevych explained. A key distinction with AIDA is that it can promote decision making based on multiple interpretations, whereas many decisions made currently rely on a single analysis that could perhaps have ignored important evidence. Allowing an end user to consider multiple possibilities and prepare for contingencies could be very useful, according to Onyshkevych.

Rama Chellappa, University of Maryland, College Park, asked Onyshkevych how confidence is measured for each modality. Onyshkevych would like to have combinable confidences across modalities but acknowledged that this is a question they have only just begun to explore. Chellappa then asked about the role of sensor fusion in AIDA. Onyshkevych responded that although AIDA will not look at structured sensors, it does build on the insight of previous sensor data exploitation. Josyula Rao, IBM, suggested that if solutions to particular conflicts depend on having a model across the different hypotheses, then multiple models of different realities could arise. Onyshkevych explained that DARPA is *allowing* rather than explicitly *enforcing* the ability to propagate multiple models. Rao added that there could be dissonance between the cognitive model that the human is working with and the model that has been programmed into the machine. Rao thus emphasized the need to converge the human–machine divide. Onyshkevych agreed that that is a non-trivial problem and noted that it may be explored in a future DARPA program. Jay Hughes, Space and Naval Warfare Systems Command, wondered how to revert back to the original source data in the hypothesis. Onyshkevych responded that no analytic product working with unstructured data will ever be 100 percent sure of anything. He continued that once one has hypotheses from earlier documents, and one is looking at the next document, these hypotheses carry some expectations, though different confidence levels may exist.

Chellappa asked if it is important to look at non-monotonic reasoning in these programs. Onyshkevych said that performers look at each piece of evidence only once before it ages out. Peter Pirolli, Institute for Human and Machine Cognition, asked whether the hypotheses are representative of what is in the document collection or if they are relevant to decisions that have utility or risk associated with them. Onyshkevych noted that those explanations are not mutually exclusive; AIDA will produce representations and hypotheses that explain the world from a particular data set. James Donlon, National Science Foundation, asked what the research community is discovering about the limits and capabilities of processing heterogeneous data sources. Onyshkevych responded that the notion of top-down expectations is something that many people rely on because trying to do purely bottom-up discovery of a model is not something people can yet do well enough. Donlon added that these DARPA programs are intriguing because they allow the performers to decide which techniques appear to be promising in new spaces.

[2] For more information about the AIDA program, see Boyan Onyshkevych, "Active Interpretation of Disparate Alternatives (AIDA)," DARPA, https://www.darpa.mil/program/active-interpretation-of-disparate-alternatives, accessed August 27, 2017.

6

Session 5: Learning from Noisy, Adversarial Inputs

OPENING REMARKS

David Honey, Office of the Director of National Intelligence

David Honey, Office of the Director of National Intelligence (ODNI), thanked everyone for participating in this experiment to change the way that ODNI does business and operations for research and development planning in its science and technology office. He stated that ODNI's main function is intelligence integration; in other words, how do the people and processes of the various agencies work together more effectively to meet national security needs?

Honey described the Intelligence Community Information Technology Enterprise (IC ITE)[1] initiative, which will move the community to a cloud-based environment where the various information systems, databases systems, and tools will coexist and promote more effective sharing. Honey initially considered this a bold step because ODNI decided to move forward with this initiative before the relevant technology had been developed. Because this approach is common in government acquisitions, he noted that smart decisions about research and development investments are crucial.

Although the intelligence community has a wide range of problems and varied members, one commonality, Honey stated, is that all of its activities revolve around collection and analysis. And collection can be complicated because the more that is collected, the larger the big data problem becomes. Thus, problems have to be considered holistically. Honey suggested that ODNI look at how the technology, the people, the processes, and the system all work together and improve the overall architecture instead of simply attempting to improve individual components. Generating partnerships and building frameworks helps ODNI to determine what research and development investments need to be made in research related to national security.

[1] For more information about IC ITE, see Office of the Director of National Intelligence, "IC IT Enterprise," fact sheet, https://www.dni.gov/files/documents/IC%20ITE%20Fact%20Sheet.pdf, accessed August 10, 2017.

HARNESSING MACHINE LEARNING FOR GLOBAL DISCOVERY AT SCALE

Mikel Rodriguez, MITRE

Mikel Rodriguez, MITRE, focused his presentation on harnessing machine learning (especially computer vision) to do global discovery at scale. He explained that because there is both an explosion of data and the potential for conflicts to arise without warning, it is imperative to develop new ways to collect intelligence. Rodriguez discussed the revolution in commercial imaging: today there are more pixels, different modalities, and more imaging companies. There is also an abundance of uncooperative and nontraditional sources (e.g., social media data) that could be harnessed to better focus our attention and our government assets, according to Rodriguez.

Although visual data may be considered the biggest form of big data, it was previously challenging to access it, index it, effectively reason about it, or move from pixels to meaning when comparing two images. These problems of representation began to be addressed with deep learning approaches, thanks to the work of Yann LeCun, and with the presence of graphical processing units.

Rodriguez commented that this thriving research field that started in academia quickly spread to industry, national laboratories, and federally funded research and development centers. Deep learning and computer vision algorithms can be used to detect images, understand patterns, and make insights about objects or events of interest. Particular attention is focused on analyzing social media data, though this is an especially noisy source, Rodriguez explained.

With the success of deep learning systems, people wonder about the role of computer vision in the future. As a community, Rodriguez suggested further investment in the areas of sensitive analytics and semantic segmentation. He noted the issue of "long tails" (i.e., the many specific or rare objects that are difficult to model because so few examples exist) in the visual world and the possibility of mixing synthetic data with real examples to create better classifiers. Rodriguez explained that labeling data is expensive: it is irrational to expect that large groups of people will be available to label all of the data the intelligence community has. Instead, it would be more effective to try to use current data sets to train algorithms.

Rodriguez said that deployment of deep learning approaches remains a challenge in certain domains. For example, deep learning approaches currently depend on expensive graphical processing units that are computationally powerful but draw much energy. While less expensive hardware could be used for object detection, Rodriguez noted that there is a balance between how much recognition power one needs and how much electrical power one is drawing from a device.

In conclusion, Rodriguez summarized the key points of his presentation: (1) Deep learning has recently led to tremendous progress in representation learning when large amounts of labeled data are available; (2) Image representation is powerful and has led to breakthroughs in classification, segmentation, and scene understanding; (3) Deep learning does not necessarily equate to deep insight; and (4) More work needs to be done on complex activities, temporal sequences, rare targets, and modalities unique to the Department of Defense/intelligence community, as well as counter artificial intelligence assurance.

Devanand Shenoy, Department of Energy, asked how often deep learning architectures are used in image processing. Rodriguez emphasized that there is much interesting work in other types of architectures, though convolutional neural networks currently work the best. He noted that neuromorphic computing is exciting but has yet to materialize. Rodriguez suggested returning to the basics to determine what should be optimized and how it should be built. Anthony Hoogs, Kitware, Inc., noted that the workshop had yet to cover the influences of deep learning and convolutional neural networks on image processing, though they have been substantial. For the intelligence community in particular, there are improved approaches for image clean-up, for example.

Josyula Rao, IBM, asked which approaches should be used when the number of images is very small and specialized and how deep learning evolves to address general problems in security. He also asked Rodriguez to provide more details on what is happening in the space of counter artificial intelligence assurance since adversaries are using artificial intelligence tools for synthesizing malware that will evade security controls. Rodriguez explained that there is a lot of work being done both on the offensive and defensive sides. He said that it is important for the intelligence community to recognize what is possible—for example, it is easy for adversaries to extract information from a model on which one has trained data. Rodriguez also remarked that there is not currently much economic incentive to work in this area, so additional investment would help to address the remaining gaps.

Pelayo Fernandez, Department of Defense, asked if synthetic data from gamers could be useful in helping algorithms learn better varieties of targets (as opposed to relying on labelers). Rodriguez commented that although there is a long history of using synthetic data in the computer vision community, he cautioned against relying on it too much, because it remains difficult to simulate the real world. Hoogs suggested that the combination of synthetic data and adversarial networks could be powerful. He also noted that the gaming community has three-dimensional models of Russian military equipment that are openly available.

Ellen Voorhees, National Institute of Standards and Technology, added that, in her experience, synthetic data does not capture the real world or the problem at hand adequately. Kathy McKeown, Columbia University, suggested that as we move to using multi-modal data, there may be cases where there are images with accompanying text that could be useful for generating captions and descriptions. Rodriguez agreed that this is an appropriate area for future research but noted the challenges that will arise between siloes that do not communicate with one another.

Rama Chellappa, University of Maryland, College Park, noted that it is difficult, if not impossible, to gain access to restricted data, though data sets for action recognition are becoming more accessible. Some sensitive data will never be available, so that challenge continues. He also explained that although synthetic data has its faults, it is still worth using as long as the limitations are understood. In reference to McKeown's earlier comment on multi-modal data, Hoogs added that there is a treasure trove of data that the intelligence community could leverage—for example, people blogging in real time while watching unmanned aerial vehicle camera footage.

7

Session 6: Learning from Social Media

LARGE-SCALE MULTI-MODAL DEEP LEARNING

Rob Fergus, New York University and Facebook

Rob Fergus, New York University and Facebook, described Facebook as a multi-modal platform with billions of images, posts, users, comments, and videos uploaded each day. Research is being conducted on a variety of deep learning approaches for handling such a heterogeneous data stream.

The first project Fergus discussed described strategies for handling multi-modal data. Using deep neural networks, World2Vec represents the many different entities found on Facebook as high-dimensional vectors in the *same space*. Fergus explained that these different types of data (e.g., users and pages) have relations between them, which are then graphed. A complete graph would have 10^{10} edges[1] per day, so this method scales to very large data streams.

Fergus noted that when analyzing these large, noisy data streams, ambiguity creates challenges that can be addressed with these methods—for example, exploring the embedding space can reveal whether a post about "football" is discussing soccer or the National Football League. These methods also allow for Facebook page categorization. Fergus summarized that large-scale embedding methods are very powerful because common space can be used to represent many different types of data, and relationships between data can be represented as distance functions within that space. He explained that these embedding methods can be considered a continuous form of reasoning.

The second project Fergus introduced described ways to train deep models from scratch and thus learn from noisy labels, such as the metadata associated with images found on Flickr. Setting up a successful architecture requires (1) training convolutional neural networks to predict words that appear with an image, (2) treating the words in the comments as targets for the photos, and (3) training convolutional neural networks to predict the words from the images. Training is done using multi-class logistic loss over all image–word pairs and mini-batch stochastic gradient descent with class-uniform sampling so that the common uses do not dominate the training.

For the experimental setup, the networks are trained and then evaluated based on word prediction, transfer learning, and word embeddings. Fergus indicated that this "weak form of supervised training" has been successful; it *is* possible to learn value representations without explicit human annotation.

---

[1] Edges are the links that connect the points of a graph.

The third program Fergus described focused on training big models faster—specifically, training ImageNet in 1 hour. He discovered that combining more data with bigger models leads to higher accuracy, although the training time also increases in such a scenario. In distributed training, as the batch size increases, optimization issues arise, so the learning rate must be scaled linearly to ensure accuracy. Increasing the number of graphical processing units also allows a linear speed-up in the training time.

The fourth and final program that Fergus presented studied training on large data sets that do not fit on a single graphical processing unit. Though the previously discussed program addressed the issue of faster training, it did not address increasing the size of the model itself. A barrier to this is that the memory of a graphical processing unit is limited. The key solution, Fergus explained, is a "hard" mixture-of-experts approach, in which each example is directed to a single expert.

Rama Chellappa, University of Maryland, College Park, noted that the mixture of experts method has also been useful in Janus's face verification research, as discussed in Chapter 3 of this proceedings. In response to a question from Kathy McKeown, Columbia University, Fergus explained that the representation could be optimized for a particular objective. Darrell Young, Facebook, asked Fergus how he would use fastText vectors between languages. Fergus thought it may be possible to do machine translation with fastText representation, though he was unsure precisely how.

8

Session 7: Humans and Machines Working Together with Big Data

SENSEMAKING SYSTEMS AND MODELS

Peter Pirolli, Institute for Human and Machine Cognition

Peter Pirolli, Institute for Human and Machine Cognition, introduced the human–computer interaction in sensemaking tasks, which are prominent in the intelligence community. He defined sensemaking tasks as tasks with massive amounts of data and with initially ill-defined goals and constraints. The purpose of completing the sensemaking task is to organize and interpret the data to generate knowledge to make a decision or provide a solution to a problem. Examples of sensemaking tasks include the need to understand a health problem in order to make a medical decision or the desire to produce an actionable intelligence report.

Pirolli continued by noting that since the cognitive capacity of human beings is relatively limited, they have to adapt to deal with this influx of available data in sensemaking tasks. Though creating more technology can be helpful, doing that alone is insufficient. The solution to this problem is to develop technology that specifically augments the rate at which humans can perform sensemaking tasks. Such system-level improvements can help humans gain more knowledge, make better inferences, and arrive at better decisions. He added that system-level improvements also have the potential to reduce cognitive bias, increase system assessments, and allow analysts to evaluate the trustworthiness of the system's products.

When human beings interact with information, Pirolli explained, phenomena occur at multiple time scales, each having their own laws. The lowest level could be considered the neurobiological (i.e., neurons fire and ensembles work together), with the next layers being the psychological (i.e., things are retrieved from memory to make a decision), the rational (i.e., the organization of tasks and their solutions), and the social (i.e., communication and organizational processes). Pirolli explained that it is important to understand how to develop predictive, multi-level models for understanding the human analysts who perform sensemaking tasks. Doing so allows (1) an understanding of the analysts' performance capabilities, biases, and failure modes; (2) an opportunity to develop better training programs or better systems; (3) a prediction of the impact of new technologies and new interaction techniques on the cognitive system; and (4) an opportunity to help artificial intelligence systems form common ground with humans to improve collaboration. The ultimate goal, he said, is to develop an interdependent joint activity between the human and the system: the user should understand what the system is doing, and the system should understand what the user is doing.

Pirolli introduced techniques developed around cognitive task analysis that try to understand how people organize their thoughts. Understanding how things are organized makes it possible to target technologies at these different levels that can impact overall performance. Information foraging theory uses mathematical models to try to understand how humans search for information and demonstrates that humans make optimizing choices. Work is also being done around highly integrated cognitive architectures to understand how memory, perception, judgment, and problem solving work together to perform tasks.

Pirolli described the Intelligence Advanced Research Projects Activity (IARPA) Integrated Cognitive–Neuroscience Architectures for Understanding Sensemaking (ICArUS)[1] program, which modeled human sensemaking and geospatial intelligence tasks in order to predict human bias and performance in such tasks. Both the task and the data are analyzed at multiple levels and mapped to a cognitive architecture, before being mapped to neural simulation models, which simulated the data to some degree of high fidelity.

Priolli's last example discussed the use of computational social science and social psychological models. Multi-level modeling is also useful in studying credibility and decision making: computational cognitive models help to better understand how people judge the credibility of others.

Current technology capabilities for sensemaking tools include (1) the use of commercial off-the-shelf interactive visual analytics (e.g., Tableau®) for big data analysis, (2) mature research platforms for visualization grammars and toolkits, and (3) early versions of interactive visualizations to understand what machine learning systems are doing. He added that a standard model of cognition has emerged for constrained, reasonably well-defined tasks. Cognitive models for tutoring systems are well established, as is user modeling and personalization.

Over the coming years, Pirolli stated, challenges will remain in human–artificial intelligence collaboration: (1) Sensemaking tasks for the intelligence community are ill-defined and dynamically changing (e.g., the data changes, as do the tactics of the adversaries); (2) Missions and tasks span multiple time scales, levels of aggregation, and levels of organization; and (3) The user and the machine will continually try to understand one another in the midst of change. Pirolli noted a Defense Advanced Research Projects Agency (DARPA) program, Explainable Artificial Intelligence (XAI),[2] that creates new learning processes and interfaces so a user can understand why a system is doing something, how to correct failures, and how to make predictions.

In the next 3–5 years, Pirolli believes the community will likely have (1) multi-level models across the cognitive, rational, and social bands for well-defined stationary sensemaking tasks; (2) an emerging field of visual analytics for machine learning programming; and (3) interactive task learning for usable soft bots for well-defined tasks. Longer-term challenges include developing (1) a foundational science of Human–Autonomy Collaboration (HAC) that would aid the transition from programming to learning to work together; (2) multi-level models of sensemaking that include dynamically adapting humans and machine learning in joint analysis; (3) research methods and evaluation frameworks on joint human–autonomy tasks; and (4) open source data and code for tasks that have some relevance and validity to the intelligence community.

Joseph Mundy, Vision Systems, Inc., highlighted the varying levels of programming skills needed for different tasks. He asked if models for intelligence analysis can predict these needs and suggested that machines could do the mundane clean-up while humans do the discovery work. Pirolli explained that there is some debate about this model for human–machine interaction. He continued that it is important to view the optimal structure as a task itself and consider the best organization between the human and the machine.

CROWDSOURCING FOR NATURAL LANUGAGE PROCESSING

Chris Callison-Burch, University of Pennsylvania

Chris Callison-Burch, University of Pennsylvania, defined crowdsourcing as hiring people to accomplish small tasks as low cost; Amazon's Mechanical Turk is the most common platform for crowdsourcing. Here, humans

[1] For information on the ICArUS program, see IARPA, "Integrated Cognitive-Neuroscience Architectures for Understanding Sensemaking (ICArUS)," https://www.iarpa.gov/index.php/research-programs/icarus/baa, accessed August 27, 2017.

[2] To learn more about the XAI program, see D. Gunning, "Explainable Artificial Intelligence (XAI)," DARPA, https://www.darpa.mil/program/explainable-artificial-intelligence, accessed August 29, 2017.

perform "artificial artificial intelligence"; in other words, humans perform artificial intelligence tasks because artificial intelligence algorithms are not yet equipped to solve certain problems.

Callison-Burch described ImageNet, a large-scale annotated resource, as a successful example of crowdsourcing that categorizes images using WordNet (a natural language processing resource) ontology. Callison-Burch highlighted a process from Fei-Fei Li that involved collecting candidate images from the internet, which were then verified by humans; over 1,000 object classes and over 1 million labeled images resulted. Li deduced that if only one person were to have to label 10,000 candidate images for each of 40,000 synsets, it would take that person nearly 20 years to complete the task. Thus Li and her team designed the ImageNet Mechanical Turk User Interface so that multiple people could work on categorizing images: This technology allowed 11 million images to be labeled by 25,000 Mechanical Turk workers in only 1 year. Although neural networks have advanced the state of the art in computer vision, a goal related to object recognition, for example, would not have been achievable without labeled training data.

Crowdsourcing has also enabled advances in natural language processing because it allows collection of data needed to train models, according to Callison-Burch. He acknowledged that while the low costs associated with crowdsourcing may be appealing, the real benefit of crowdsourcing is that annotation tasks can be completed quickly by a large number of people.

Callison-Burch explained that machine translation takes a sentence in one language and produces a translation in another language. These translation systems are typically constructed using paired sentences (i.e., parallel data) in which a human has engaged in a translation task. . Certain languages that are of interest to the intelligence community, however, often have very little parallel, or training, data available. Machine translation could be improved, according to Callison-Burch, either by building a better model that accomplishes the task and performs better or by adding more data to the training collection.

During his work with the Human Language Technology Center of Excellence, Callison-Burch quantified the translations from Mechanical Turkers. He found that while machine translations are similar to Turker translations, both are worse than those of professional translators. With additional quality control, he found that he could improve the quality of the output from the Mechanical Turkers.

Callison-Burch performed a number of experiments to study the capabilities of human language technologies on Mechanical Turk. He researched how to create bilingual parallel text for a variety of low resource languages; completed a large-scale demographic study in which the list of languages represented by Mechanical Turk workers increased; and used crowdsourcing to identify cases of written Arabic dialect and create an Arabic-to-English parallel corpus that could be used to train a machine translation system.

He explained that there is a need for low-cost, high-quality translations, especially to expand the number of languages currently covered by Google and the Center for Applied Machine Translation. There are both advantages and disadvantages of crowdsourcing, he reiterated: while it is scalable, costs little, and offers an on-demand workforce, there are quality concerns, it is impossible to make assumptions about skill levels, and it cannot be used for sensitive data. In conclusion, he emphasized that crowdsourcing is a powerful tool that is worthy of additional investment, as it helps enable better data creation and thus better science and technology. Within the next 5 years, Callison-Burch expects to see a tighter integration of crowdsourcing and machine learning (e.g., active learning or domain adaptation) as well as a new crowdsourcing platform for natural language processing that could be deployed on the inside for the intelligence community.

Rama Chellappa, University of Maryland, College Park, asked if there is any research on Turkers who do both image and natural language processing work. Callison-Burch said that work at the intersection of language and vision is only just beginning and added that different language capabilities allow for different tasks. Jonathan Fiscus, National Institute of Standards and Technology, mentioned a group at the National Geospatial-Intelligence Agency that is currently building a new crowdsourcing platform. Kathy McKeown, Columbia University, asked if Callison-Burch could discuss overall systems that had components that would sometimes be machine and sometimes human. He responded that the crowd component is usually placed at the start of the pipeline; however, it would be beneficial to have a component in a live pipeline that made a decision in real time about whether the predictions being made by the machine learning algorithms were good enough, and, if so, the work could be routed there. If not, the work could be routed to the crowd instead.

9

Session 8: Use of Machine Learning for Privacy Ethics

TOWARD SOCIO-CULTURAL MACHINE LEARNING

Mark Riedl, Georgia Institute of Technology

Mark Riedl, Georgia Institute of Technology, explained that researchers want to develop artificial intelligence systems that can learn about societies and cultures and then act more ethically safe around humans. He made a distinction between artificial intelligence ethics and ethical artificial intelligence: Artificial intelligence ethics concerns the proper use of artificial intelligence technologies in society, while ethical artificial intelligence considers how to make systems safer to use. According to Riedl, machines will need to learn the culturally appropriate way to make decisions.

Riedl next discussed the issue of value alignment: intelligent, autonomous agents need to understand how the culture works so as to pursue only those goals that are beneficial to humans. There are, however, two perspectives of value alignment. The strong form of value alignment, which constrains robots from intentionally or unintentionally performing behaviors that would harm humans, seems nearly impossible to Riedl. The weak form of value alignment, which Riedl described as more reasonable, creates intelligent systems to share our values and perform behaviors that conform to them.

Riedl presented Asimov's three laws of robotics[1] as a way to address this issue of value alignment but explained that these laws would be insufficient because they offer no clear definition of "harm." (i.e., Is harm physical, psychological, or both?) Because it is impossible to use hard-coded restraints for ethics, Riedl proposed an alternative way to address value alignment for artificial intelligence—reinforcement learning involves trial and error in a simulated or real environment and rewards an agent's appropriate behavior accordingly. Using reinforcement learning requires that humans be very specific in their communications with robots; unfortunately, humans are generally not good at saying *exactly* what they mean. For example, if one tells a robot to pick up medication from the store, it will do just that (without paying for it). In addition to these types of sensor errors, he noted, there can also be motor errors, problems with long-term look-ahead, and corrupted rewards when deploying reinforcement learning on artificial intelligence. Reinforcement learning agents learn from feedback, demonstration, or stories, but they only learn about as much of the environment as they *need* to in order to achieve the rewards (see Figure 9.1).

[1] For a description of Asimov's laws, see The List of Lists, "Isaac Asimov's 'Three Laws of Robotics,'" https://www.auburn.edu/~vestmon/robotics.html, accessed September 4, 2017.

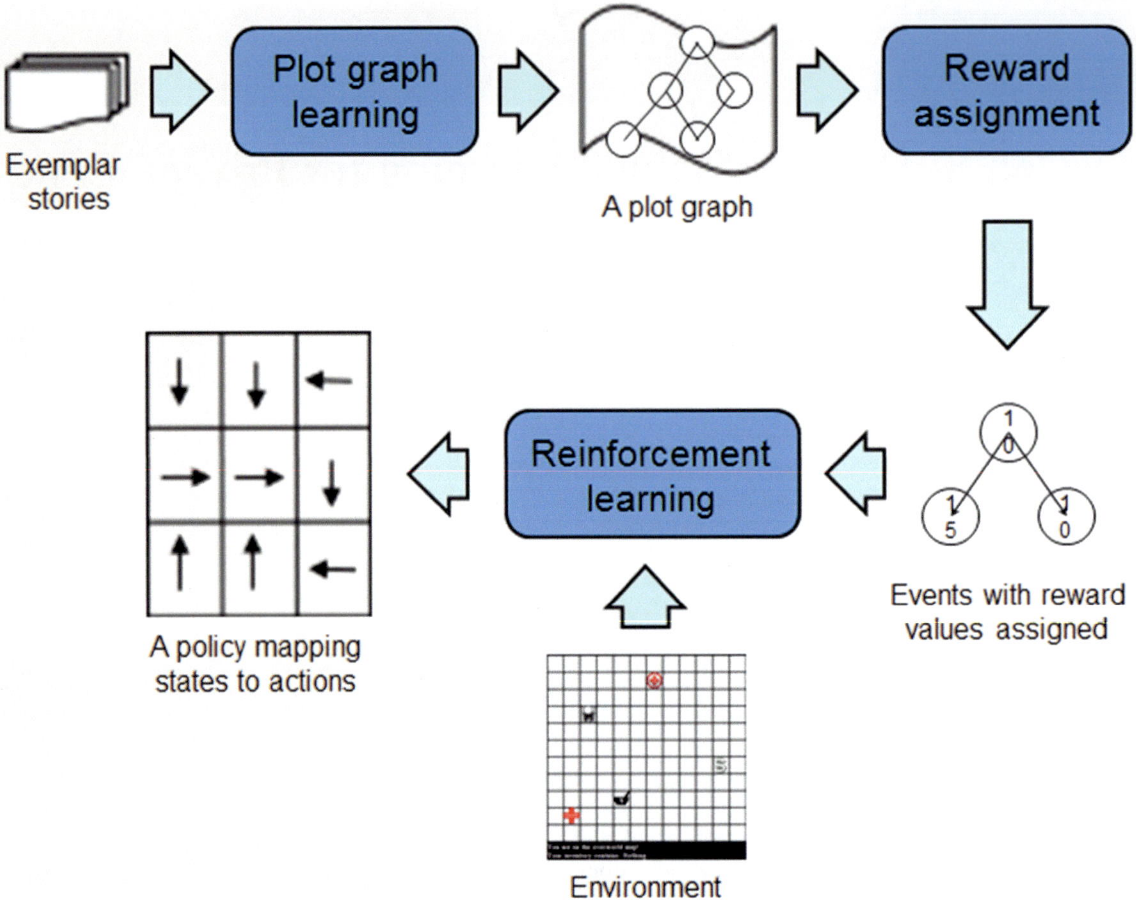

FIGURE 9.1 Reinforcement learning from stories. SOURCE: Mark Riedl, presentation to the workshop, August 10, 2017.

Riedl explained that the next challenge in creating ethical artificial intelligence is the question of how robots learn the rules of society: learning a little bit about everything is a daunting task, however. The rules of the world, often referred to as "commonsense knowledge," can be either declarative or procedural. Procedural commonsense knowledge includes sociocultural conventions that reduce human–human conflict. Riedl pointed out that databases are being created to collect commonsense knowledge; however, declarative knowledge databases are missing information and, of course, contain no procedural information. One suggestion he mentioned is to put sensors everywhere in the world to collect video and then parse it to collect an infinite amount of information, but such an approach is neither realistic nor desirable.

Yet, Reidl noted, machines will still be expected to understand patterns of human behavior ("machine enculturation") far beyond basic activity recognition. Rapid innovation continues in computational imagination by combining context with expectation to generate a prediction, mimicking humans' ability to make inferences. Machine enculturation can look at issues in commonsense failure goals: if a robot does not complete a task in the exact way the human intended, is that considered a failure since the robot executed on the wrong goal? Machine enculturation can also be used to train humans to understand social conventions. In addition to reinforcement learning, he stated, videos and images can also be used to learn commonsense knowledge and machine enculturation. Human cultural values are implicitly encoded in stories told by members of a culture—for example, through allegories, fables, literature, television, and movies.

Riedl defined "curated corpora" as data that have been selected because they all relate to the same topic. Though it is difficult to create such a data set, it can be useful in teaching systems about common sociocultural conventions. When the data are curated, it is possible to learn some sort of model and see temporal relations between events (i.e., there are things that happen regularly, and they happen in a particular order). He noted that having a curated corpora can present a bottleneck, however, and deep learning is needed to learn human values/conventions from stories on the Internet. But since what happens next in a story is not always what actually happens next chronologically, a sparsity problem then results. Riedl emphasized that different agents will be needed for different cultures, so a question also arises about who will decide what the artificial intelligence sees, reads, and learns. And to improve predictions, Riedl noted that machine vision can be used alongside stories.

Researchers in the field are currently studying problems that will likely not emerge for another 5 or more years. Riedl noted that systems that are aware of sociocultural conventions or curated data sets do not currently exist, and researchers still do not know how to solve these problems using deep neural networks. Three to 5 years from now, Riedl expects agents to start using commonsense knowledge and world knowledge to address human needs; to engage in longer conversations; to rely on computational imagination; and to differentiate behavior based on cultural context. Ten years from now, he hopes to see an increasing presence of autonomous systems in social contexts (although the problem of humans understanding artificial intelligence and artificial intelligence understanding humans will be far from solved) as well as ways to address broader artificial intelligence safety concerns. And in 10–20 years, the use of military teammates could become more common.

Joseph Mundy, Vision Systems, Inc., asked if Google and Facebook are doing this type of research since systems such as Siri and Echo would need similar information for their tasks. Riedl acknowledged that these larger companies are starting to think about this in conjunction with academic researchers. Academic research is hopefully helping these organizations to develop a path forward. Rama Chellappa, University of Maryland, College Park, asked if the robot or the human should take responsibility for the illegal activity that resulted in Riedl's example when the robot picked up the medication for the human without paying for it. Riedl explained that a question like this is precisely why the legal system needs to play a role in artificial intelligence.

10

Session 9: Evaluation of Machine-Generated Products

Anthony Hoogs, Kitware, Inc.
Jonathan Fiscus, National Institute of Standards and Technology
Rob Fergus, New York University and Facebook
Jason Duncan, MITRE Corporation

After each panel participant provided an overview of his work, the panel discussion focused on how to couple evaluation[1] with annotation[2] to achieve maximum performance. Annotation is a critical component for deep learning in order to generate sufficient data independent of evaluation.

Anthony Hoogs's company, Kitware, Inc., is the test and evaluation support contractor for the Intelligence Advanced Research Projects Activity (IARPA) Deep Intermodal Video Analytics (DIVA) program, a 5-year project for ground surveillance video analytics. DIVA's objective is to dramatically advance the state of the art in (1) automatic detection of activities in cluttered scenes, (2) temporal reasoning of video for improved activity detection, and (3) activity detection and scene understanding from overlapping and non-overlapping camera viewpoints. Kitware's role in achieving this objective is to collect and annotate data, develop an open source video analytics framework, and establish a baseline algorithm. Kitware was chosen primarily because of the surveillance video data it collected (and kept sequestered) during a prior program. So that the restricted data can be used for deep learning, test, and evaluation, for DIVA as well as for other programs, it needs to be annotated; Hoogs explained that it was relatively easy and cost-effective to hire 8 part-time in-house annotators to complete this task.

Jonathan Fiscus, National Institute of Standards and Technology (NIST), shared lessons learned about the qualities of a good evaluation. Evaluations are driven by many factors—for example, task design, data, infrastructure, community engagement, and testing and analysis. He added that evaluation can support transition, if done correctly. In particular, he described the following:

- *Task design* requires an independent evaluator as well as clear technology use cases and purpose. He highlighted the importance of collaborating to convert the statement of what technology needs to do into

[1] The purpose of evaluation is to "provide testing conditions that convincingly suggest that [an] algorithm will perform well on other people's data, out in the real world. . . . [I]t's important to keep track of the testing conditions, any modifications [made] to [the] algorithm, and places in [the] annotation scheme that could be changed to improve performance later" (A. Stubbs and J. Pustejovsky, 2013, *Natural Language Annotation for Machine Learning: A Corpus-Building for Applications*, O'Reilly Media, Sebastopol, Calif.).

[2] Annotation is described as "any metadata tag used to mark up elements of a data set. . . . [I]n order for the algorithms to learn efficiently and effectively, the annotation done on the data must be accurate and relevant to the task the machine is being asked to perform" (A. Stubbs and J. Pustejovsky, 2013, *Natural Language Annotation for Machine Learning: A Corpus-Building for Applications*, O'Reilly Media, Sebastopol, Calif.).

evaluation tasks. Task design can be innovative or prescriptive, and the evaluation can be component or end-to-end.
- *Data* is needed for training, development testing, cross validation, and evaluation testing. He emphasized the need to have enough data to test system differences and to divide up evaluation resources into publicly available and sequestered data.
- *Infrastructure* impacts transition, and innovative programs can be more difficult to transition than prescriptive programs. He suggested using the cloud for effective testing.
- *Community engagement* can be a challenge when difficult problems have competing requirements. It is crucial that evaluators understand the relevant trade-offs.
- *Testing and analysis* is key for transition. Unfortunately, there is no way to automatically evaluate a new machine learning algorithm. Possible solutions to this problem include task-adaptable annotations.

Fiscus discussed the Defense Advanced Research Projects Agency (DARPA) Media Forensics (MediFor)[3] program to highlight the challenges of doing so many evaluation tasks with only one data set. He noted that his team is designing data sets and evaluation tasks to do better factor analysis, which helps to decide the next steps in programming, evaluation, and research.

Rob Fergus, New York University and Facebook, described the "sweet spot" for the amount of data that can be used for deep learning. With too little data, it is difficult to train existing deep learning models; with too much data, it is difficult not to become overwhelmed. And when graduate students try to run big models on only one or two graphical processing units, the number of experiments that can be run decreases. This issue became apparent in a 2010–2011 DARPA deep learning program when graphical processing units did not have as much power as they have now. Fergus explained that this group would have benefitted from appropriate infrastructure and hardware to handle that much data. However, hardware of this caliber is quite expensive, especially for use by university students, and many universities are ill-equipped for large-scale graphical processing unit deployments. From an evaluation perspective, Fergus noted that when developing algorithms, it is essential to cycle many times and complete hundreds or thousands of experiments.

Jason Duncan, MITRE, emphasized that an important goal associated with evaluating machine-generated products is to maintain situational awareness. Researchers want to learn about the world; doing so enables strategic warning and predictive scenarios and helps intelligence personnel get ahead of some threats before they occur. Duncan noted that both technology evaluations (e.g., core technologies) and utility evaluations (e.g., value to mission) are useful in achieving situational awareness. He described the value of multi-source data in knowledge discovery and the benefits of combining evaluation metrics across data input types. And he noted that when dealing with heterogeneous data with ambiguity, it is important to measure the system's ability to have confidence in its predictions. He added that noise is inevitable in the output of unstructured content processing systems, but in many cases utility evaluations can demonstrate the value to mission of such systems even in the face of imperfect system output.

James Donlon, National Science Foundation, asked if graduate students have enough computing power to make better use of data for academic research and keep up with industry contributions. Fergus responded that the answer to this question depends upon the university in question—some of the larger universities have access to more graphical processing units. He suspects that image work is manageable on most campuses, while research in video remains a difficult computational problem. Hoogs added that the alignment of models to graphical processing units also needs to be considered: even if a university has multiple graphical processing units, this may be inadequate for a model that needs to fit into the *memory* of one graphical processing unit. Donlon asked if the ability to discover is then hindered by researchers' need to conform. Fergus responded in the affirmative and explained that conservative solutions are likely if one is limited to running only a small number of experiments. And in reference to Hoogs' point about model alignment, Fergus noted that most people do many different experiments in parallel instead of completing a single job across multiple graphical processing units.

---

[3] To learn more about the MediFor program, see D. Doermann, "Media Forensics (MediFor)," DARPA, https://www.darpa.mil/program/media-forensics, accessed August 29, 2017.

Hoogs referenced MediFor as a program that successfully provides computing resources. Fiscus added that while that is a great solution, a technology problem arises that can be addressed with cloud architecture. Hoogs added another downside to using program-wide compute resources as MediFor is doing: while shared resources work well for development, they may not be as useful when an evaluation deadline is present and all performers need significant computing resources at the same time. This, then, is a program problem, not just a performer problem. Fiscus noted that his group is currently collecting algorithms using MediFor, but it is too early to gauge whether it is successful.

Rama Chellappa, University of Maryland, College Park, asked if Facebook would be willing to develop agreements to work with academic professors and researchers, instead of only recruiting them. Fergus said that there could be security issues that would prevent such an arrangement, but discussions about the possibility continue at Facebook. Fergus noted that Amazon already has the infrastructure to do something similar: its Elastic Cloud Compute[4] could give credits to university researchers. An audience participant added that, in addition to offering academic credits to researchers, Amazon offers an academic rewards program and funding for Ph.D. students.

Donlon asked the panel about the role of reference data sets; ImageNet is central to academic research, and yet reference data sets can also constrain the machinery and capabilities of that machinery. Donlon continued that this issue is relevant for a customer who wants to deploy systems in the real world and asked how to use these reference data sets to guide academic research forward. Duncan responded that the advantages outweigh the disadvantages for reference data sets; especially in the case of government programs, reference data sets help to focus performers on the problem of interest. As an example, Hoogs said that he could not recall a major computer vision program sponsored by the USG that had full training and testing data that was completely open and useful to the computer vision community. Hoogs added that any time a program can set up a full, complete data set, it should, so as to engage the entire community. Fergus explained that one of the reasons academics often conform to reference data sets is because they have a vested interest in publishing papers. Making evaluation results and data sets available to the public adds credibility and gives people an opportunity to compare their work. Duncan added that much of this discussion is dependent upon the problem at hand for the performers.

Kathy Lajoie Malik, Technology Leadership Achieving Intended Outcomes, asked if there has been enough discussion of evaluating the ability to manage change on both the product and the processes, now that the time between research and implementation is shortening. Duncan responded that technology transfer has always been and will continue to be a difficult problem, especially since many analysts are resistant to change. Fiscus added that part of that problem could be addressed in the technology design and in evaluating the technology itself. Hoogs emphasized that it is critical to have the potential technology transition groups and end users represented in the construction of the evaluation.

[4] The website for Elastic Cloud Compute is https://aws.amazon.com/ec2/, accessed September 4, 2017.

11

Session 10: Capability Technology Matrix

MACHINE LEARNING FOR ENERGY APPLICATIONS

Devanand Shenoy, Department of Energy

Devanand Shenoy, Department of Energy (DOE), explained that although DOE is not currently supporting any artificial intelligence or machine learning projects, there is a lot of investment from industry in these areas, and DOE is interested in relevant energy applications. Shenoy provided an overview of the offices within DOE, with particular attention to the Energy Efficiency and Renewable Energy sector, and noted that DOE has oversight of over 17 national laboratories. He also provided an overview of the U.S. energy system.

Machine learning has interesting applications for renewable energy in particular, according to Shenoy. Trained on data from 1,600 U.S. sites, IBM's Watt-Sun is a self-learning weather model and renewable forecasting technology that is more accurate than previous solar forecasting models. He added that machine learning also has applications for energy efficiency: it can be applied to cooling systems data to achieve energy reduction in data centers as demonstrated by Google. Artificial intelligence can be used to minimize wasted energy and balance the electric grid by exploiting energy demand based on analytics collected from sensors. Shenoy added that smart wells that can sense temperature, pressure, chemicals, and vibrations can be deployed to streamline efficiency and mitigate failures in the oil and gas industry.

Shenoy described an occupant tracking and building energy management technology emerging from Carnegie Mellon University that utilizes floor vibration to better monitor who is in a building at a particular time. He also described structural health monitoring systems that use sensors and pattern recognition to predict future structural damage in bridges and buildings.

To learn more about opportunities for machine learning, Shenoy encouraged participants to review the Quadrennial Technology Review, which contains detailed energy technology assessments. Shenoy described some additional areas that could benefit from machine learning, many of which cross multiple industries:

- *Critical materials*. Reduce the reliance on imported materials and increase recycling of materials.
- *Sustainable manufacturing*. Provide the opportunity to design materials that can be reused or recycled, thus reducing energy usage.
- *Combined heat and power.* Combine heat and electricity more efficiently.
- *Waste heat recovery.* Recover waste heat, which in turn presents cost-saving opportunities.

- *Advanced sensors, controls, platforms, and modeling for manufacturing.* Improve predictive maintenance and product customization. This is the topic area for one of the Manufacturing Institutes supported through the Advanced Manufacturing Office at DOE. There are currently 14 Manufacturing Institutes supported through the Department of Defense (DOD), DOE, and the Department of Commerce. These Institutes typically have $70 million of funding from the government, with at least a 50:50 cost share from industry, state and local governments, and other private entities.
- *Process heating.* Offer new ways to save energy and increase efficiency by heating materials in different ways.
- *Process intensification.* Scale down the size of chemical factories while increasing efficiency.
- *Roll-to-roll processing.* Reduce the cost to print electronics and other functional materials.
- *Composite materials.* Improve technologies in lightweighting and thermosetting.
- *Additive manufacturing, advanced materials manufacturing, and materials for harsh service conditions.* Innovate to improve efficiency and to extend life of materials.
- *Wide bandgap power electronics.* Increase efficiency by replacing silicon and converting power (for example, direct current to alternating current) using silicon carbide and gallium nitride materials.
- *Thermoelectric and direct energy.* Convert heat into useful power to be more efficient in the overall system.

Shenoy turned to a discussion of challenges for microelectronics, including the concern that the electricity supply expected to be used for microelectronics will soon be unsustainable due to current transistor scaling constraints. To avoid this potential crisis, the efficiency improvement rates of the transistors and the data interconnects would need to double, according to Shenoy. Several DOE laboratories are collaborating on a multi-scale co-design framework to help address this challenge. Three DOE offices are also joining forces to address this challenge. Shenoy concluded by emphasizing that one of the best ways to address national security challenges, economic challenges, and energy efficiency challenges is through building public-private partnerships.

Jonathan Fiscus, National Institute of Standards and Technology, asked if DOE can create data sets to facilitate the kind of research discussed in Shenoy's presentation; if so, they could be used for challenge competitions. Shenoy responded that DOE's Advanced Manufacturing Office recently held a workshop on "Artificial Intelligence Applied to Materials Discovery and Design," and the information gleaned from that workshop could inform a potential program to generate such data. In response to a question from Lih Young, previous candidate for U.S. Senate, Maryland, Shenoy mentioned that the Defense Advanced Research Projects Agency is studying the value of public-private partnerships for microelectronics.

USING METROLOGY TO IMPROVE ACCESS TO "UNSTRUCTURED" DATA

Ellen Voorhees, National Institute of Standards and Technology

Ellen Voorhees, National Institute of Standards and Technology (NIST), shared NIST's mission to promote U.S. innovation and industrial competitiveness by advancing measurement science, standards, and technology in ways that enhance economic security and improve our quality of life. She noted that because it is not a regulatory body and its primary mission is not funding research, NIST is a unique federal government entity. NIST is involved in evaluation because it offers a technology-neutral site with both technical understanding and expertise in developing evaluations. NIST also keeps track of all of the artifacts via a public proceedings, data sets (often publicly available), and an archive of past results.

Voorhees explained that NIST's work in artificial intelligence spans its broad portfolio. NIST's long-standing program in metrology for information access creates evaluation infrastructure through community evaluations, the point of which is to support research and thus improve the technology. NIST's portfolio focuses on diverse sources (e.g., speech, text, images, video, biometrics) and diverse tasks (e.g., recognition, search, extraction, summarization), and NIST has worked in close collaboration with the intelligence community for at least the past three decades.

Voorhees emphasized that community evaluations form and/or solidify a research community, which helps to make progress on problems, establish research methodology, facilitate technology transfer, document the state of the art, and amortize the costs of infrastructure. She noted, however, that there are some drawbacks of community evaluations, including that they can take money and time from other efforts and they can have a problem with overfitting data sets.

Voorhees defined a good evaluation task as an abstraction of a real-world task that is controllable enough for the researcher to understand what is causing the performance and realistic enough that it captures the important aspects of the task. She added that evaluation metrics must predict the relative effectiveness of systems on the real task. The evaluation task must also have an adequate level of difficulty so that something can be learned. She noted that it is best if measures used to score the results of an evaluation are diagnostic. Voorhees concluded her talk with a description of the many different tasks and document types evaluated through the Text REtrieval Conference (TREC).[1]

Anthony Hoogs, Kitware, Inc., noted that, historically, the computer vision community has not used the data sets from TRECVid because of the associated competition constraints and data restrictions, despite the fact that it began as the largest, most formalized evaluation in the community. He highlighted a number of other competitions that do not have similar barriers. Voorhees explained that the restrictions Hoogs described do not occur with all tracks of TRECVid, and she noted that the expectation of the annualized challenge cycle is not that people will participate every year.

Peter Pirolli, Institute for Human and Machine Cognition, wondered if it would make more sense to spend time trying to understand the interaction between machines and people and to understand more about the tasks themselves than to focus on evaluating system metrics. Voorhees said that NIST's focus on evaluating systems is informative and valuable because the community learns important information about technology, though she acknowledged that other challenges and open research questions remain.

CHALLENGE PROBLEMS FOR MULTI-SOURCE INSIGHTS

Travis W. Axtell, Office of the Under Secretary of Defense for Intelligence

Travis Axtell, Office of the Under Secretary of Defense for Intelligence, shared a sampling of comments from various DOD researchers. One researcher's description of machine-augmented analysis of multi-source data is that multi-source apertures are intrinsically concurrent and semi-sequential (i.e., data sources are generally processed with machines using multiple execution flows that progress simultaneously [intrinsically concurrent], with a staggered offset between flows [too large of an offset yields sequential execution]; the semi-sequential nature of processing does add some processing latency or delay of resulting computation). They added that while latency is often a barrier to progress, it can be beneficial if the result is improved accuracy. They noted the importance of time synchronization across sources and that components from various elements of the data-to-decision chain should weave together in an interoperable framework. They highlighted the value of data-driven insights from and across multiple scales.

Axtell shared the results from an Air Force Research Laboratory experiment to introduce a topic of value for future research: training an algorithm to understand when the change detection aspects are working at a practical consideration for an operator. Another area worthy of study is the parallel development in intelligence community organizations of processing frameworks. Axtell next explained that because algorithms used today have difficulty registering moving objects, moving objects could be a focus of multi-source research. He warned that recognition and registration of moving objects can be challenging, but registration aspects will lead to interesting developments in the future. Axtell explained that it is also important to be able to share information from one source with another. Eventually, he hopes that single-source processing and multi-source processing can be automated with machine learning.

Axtell explained that there are a number of ways to set up and solve an optimization problem. He mentioned

[1] For information about TREC, see http://trec.nist.gov, accessed August 29, 2017.

YALMIP,[2] a MATLAB toolkit that symbolically represents an optimization problem and then converts the problem to match the capabilities of the available optimization solver. According to Axtell, the intelligence community would benefit from having a symbolic representation of the neural network's intent. With this representation, the computer could then propose an appropriate design and present a solution.

Hoogs mentioned that time synchronization can be a problem when multiple video sources are looking at the same outdoor scene and one wants to know the three-dimensional structure of an object in that scene—cameras often do not have an exact notion of time to be able to synchronize down to the frame level. Axtell responded that the capability exists to develop the registration and solve for that registration between those two sources, which is more fruitful than trying to place clocks in the scenes. Hoogs added that latency (i.e., the time between capturing the image and reporting the element) and computational costs are also problematic in the computer vision community, especially *with* advanced methods and graphical processing units. Axtell noted that uncertainty is also a factor in these issues.

In response to a question from an audience participant, Axtell explained that a time representation is needed to infer a relationship between two data files. In response to another question from the audience, Axtell emphasized the importance of understanding the conditions of each sensor type. He added that with higher speed comes more data, which makes the task of knowledge discovery even more difficult.

AN OVERVIEW OF NATIONAL SCIENCE FOUNDATION RESEARCH IN DATA ANALYTICS

James Donlon, National Science Foundation

James Donlon, National Science Foundation (NSF), summarized NSF's mission as transforming the frontiers of science and engineering. NSF funds approximately 12,000 of the 50,000 proposals it receives each year. In FY 2017, approximately $7.5 billion was allocated to support fundamental scientific research. NSF's Directorate for Computer and Information Science and Engineering (CISE),[3] which supports machine learning, artificial intelligence, and data analytics research, received $934 million in FY 2016. Overall, NSF is responsible for 82 percent of the federal support of academic basic research in computer science. He emphasized that although basic research may *seem* like it is not pointed enough to meet customer needs, without it, it is impossible to make larger discoveries.

Donlon characterized NSF's research portfolio, with particular attention to machine-augmented analysis of multi-source data. CISE has three core programs in its division on Information and Intelligent Systems, which supports research and education activities that study the interrelated roles of people, computers, and information. The Robust Intelligence program's goal is to advance and integrate the traditions of artificial intelligence, computer vision, human language research, robotics, machine learning, computational neuroscience, cognitive science, and related areas. The Information Integration and Informatics program studies the creation, management, visualization, and understanding of diverse digital content, while the Cyber-Human Systems program studies issues of user interface, assistive technology, human–computer interaction, and the combination of social media and social networks. Donlon encouraged workshop participants to visit NSF's website and use its publicly accessible "NSF Award Search" feature[4] to search for the many active awards in this research space.

To understand the themes that play the largest role in the research supported by NSF, Donlon performed a cluster analysis of titles and abstracts of awards. The analysis showed machine learning to be the most dominant research theme of the Robust Intelligence and Information Integration and Informatics divisions; however, this only means that the phrase is used often as a label, not necessarily that this theme plays a central role in research. A list of recent grants includes topics in foundational machine learning—for example, adversarial machine learning, deep reinforcement learning, learning structured prediction models, uncertainty representation, and personalizing from observational data. And a select list of active awards includes multi-modal research in computer vision and

[2] The website for YALMIP is https://yalmip.github.io, accessed August 29, 2017.

[3] The CISE website is https://www.nsf.gov/dir/index.jsp?org=CISE, accessed August 29, 2017.

[4] The website for NSF's Award Search is https://www.nsf.gov/awardsearch/, accessed August 29, 2017.

human language technology, although Donlon noted from his search that, generally, problems of multi-modality seem to be of greater interest to practitioners than to NSF-supported researchers. He added that it is important for funding agencies to think about how to better encourage and fund multidisciplinary research.

CISE's "national priorities" include big data, cybersecurity, the National Robotics Initiative, understanding the brain, the National Strategic Computing Initiative, Smart Cities, Computer Science for All, and advanced wireless research. Donlon concluded his talk with a summary of what he discovered about NSF's funding in relation to the themes of the workshop:

- Machine learning is everywhere and has been especially successful within a single modality on data-rich classification problems and where the problem space is sufficiently constrained.
- Multi-modal, multi-source data was a focus of only a small number of projects, which indicated that the funding agencies may have more work to do in engaging the community.
- Solving the end-to-end problem will always be an engineering endeavor that brings to bear outcomes from research across the whole spectrum.
- The impact of big data is evident everywhere, as are the problems and opportunities for social networks.

Kathy McKeown, Columbia University, agreed with Donlon that NSF needs to think carefully about how it funds interdisciplinary research. She also noted that it is very difficult to secure funding for big data research if it does not immediately connect to the foundational area that is being supported. She suggested NSF consider creating another core program that would be more interdisciplinary in nature to address both of these funding issues. Donlon acknowledged McKeown's input as very useful for NSF. He introduced NSF's Research Advanced by Interdisciplinary Science and Engineering (RAISE) program, which invites multidisciplinary research and reduces some of the stresses on the merit review community, but he noted that NSF has more work to address this systemic problem.

Hoogs highlighted the new field of visual question answering (i.e., an example of integration of vision and language) that plays a prominent role in the computer vision community. He suggested that it would be valuable to look at where the experts in this field received their funding since they were able to bridge the gap between fields successfully.

Appendixes

A

Biographical Sketches of Workshop Planning Committee

RAMA CHELLAPPA, *Chair*, is a distinguished university professor and a Minta Martin Professor of Engineering and chair of the ECE department at the University of Maryland. Prof. Chellappa is a recipient of an NSF Presidential Young Investigator Award and four IBM Faculty Development Awards. Prof. Chellappa received the K.S. Fu Prize from the International Association of Pattern Recognition (IAPR). He is a recipient of the Society, Technical Achievement, and Meritorious Service Awards from the IEEE Signal Processing Society. He also received the Technical Achievement and Meritorious Service Awards from the IEEE Computer Society. At UMD, he received college and university level recognitions for research, teaching, innovation, and mentoring of undergraduate students. In 2010, he was recognized as an Outstanding ECE by Purdue University. He received the Distinguished Alumni Award from the Indian Institute of Science in 2016. Recently, he received the inaugural Leadership Award from the IEEE Biometrics Council. Prof. Chellappa served as the editor-in-chief of PAMI. He is a Golden Core Member of the IEEE Computer Society, served as a Distinguished Lecturer of the IEEE Signal Processing Society and as the president of IEEE Biometrics Council. He is a fellow of IEEE, IAPR, OSA, AAAS, ACM, and AAAI and holds six patents. His current research interests span many areas in image processing, computer vision, machine learning, and pattern recognition.

THOMAS DIETTERICH (A.B., Oberlin College, 1977; M.S., University of Illinois, 1979; Ph.D., Stanford University, 1984) is professor emeritus and director of Intelligent Systems Research in the School of Electrical Engineering and Computer Science at Oregon State University, where he joined the faculty in 1985. Dietterich has dedicated his career to machine learning and artificial intelligence. He has authored more than 180 publications and 2 books. His research is motivated by challenging real-world problems ranging from personal information management, to drug design, to sustainability, and most recently to problems in safe and robust artificial intelligence. Dietterich has also devoted many years of service to the research community. He is past president of the Association for the Advancement of Artificial Intelligence, and he previously served as president of AAAI (2014-2016) and as the founding president of the International Machine Learning Society (2001-2008). Other major roles include executive editor of the journal *Machine Learning* (1992-1998), co-founder of the *Journal for Machine Learning Research* (2000), and program chair of AAAI 1990 and NIPS 2000. He is currently the moderator for machine learning on arXiv. Dietterich is a fellow of the ACM, AAAS, and AAAI.

ANTHONY HOOGS is the senior director of computer vision at Kitware, a small scientific computing R&D firm based on open source software. Dr. Hoogs leads Kitware's computer vision group, which he started in 2007 and now has 35 members including 15 PhDs. For more than two decades, he has supervised and performed research in various areas of computer vision including event, activity, and behavior recognition; motion pattern learning and anomaly detection; tracking; remote sensing imagery analytics; image segmentation; object recognition; and content-based retrieval. At Kitware he has initiated and led more than two dozen projects in image and video analysis, involving more than 20 universities and sponsored by government institutions including DARPA, AFRL, IARPA, ONR, OSD, NGA, and others, ranging from basic academic research to developing advanced prototypes and demonstrations installed at operational facilities. Previously at GE Global Research (1998-2007), Dr. Hoogs led a team of vision researchers on projects sponsored by the U.S. Government, Lockheed Martin, and NBC Universal. He has published more than 80 papers in computer vision and has served as general co-chair for the IEEE Conference on Computer Vision and Pattern Recognition (CVPR) 2017; general co-chair for the IEEE Winter Conference on Applications of Computer Vision (WACV) 2016 and 2018; area chair for CVPR (2009, 2010, 2012, 2018); Workshops co-chair for CVPR (2012); corporate relations chair for CVPR (2009, 2010) and the International Conference on Computer Vision (2013); program co-chair for WACV (2009, 2011); organizer for various CVPR and ICCV workshops; member of the Computer Vision Foundation Advisory Board and Industrial Advisory Board; member of the steering committee for WACV. He has served on technical panels for NSF and DARPA, including DARPA Information Science and Technology (ISAT) panels in 2007, 2009, and 2013 and has recently started a 3-year term as an ISAT member. In 2014 he served as an organizer of the National Academies' National Research Council Workshop on Robust Methods for the Analysis of Images and Videos for Fisheries Stock Assessment, sponsored by NOAA, then joined the NOAA Automated Imagery Analysis Strategic Initiative Steering Committee. Dr. Hoogs received a Ph.D. in computer and information science from the University of Pennsylvania in 1998; an M.S. from the University of Illinois, Urbana-Champaign in 1991; and a B.A. magna cum laude from Amherst College in 1989.

JOHN E. KELLY III (NAE) is IBM senior vice president, Cognitive Solutions and IBM Research. Dr. John E. Kelly III is focused on the company's investments in several of the fastest-growing and most strategic parts of the information technology market. His portfolio includes IBM Analytics, IBM Commerce, IBM Security, and IBM Watson, as well as IBM Research and the company's Intellectual Property team. He also oversees the development of units devoted to serving clients in specific industries, beginning with the April 2015 launch of IBM Watson Health. In this role, Dr. Kelly's top priorities are to stimulate innovation in key areas of information technology and to bring those innovations into the marketplace quickly; to apply these innovations to help IBM clients succeed; and to identify and nurture new and future areas for investment and growth. He works closely with the leaders of these units to drive business development, accelerate technology transfer, and invest for the future. Dr. Kelly was most recently senior vice president and director of IBM Research, only the tenth person to hold that position over the past seven decades. Under Dr. Kelly, IBM Research expanded its global footprint by adding four new labs (including IBM's first in Africa, South America, and Australia), creating a network of approximately 3,000 scientists and technical employees across 12 laboratories in 10 countries. During his tenure, IBM maintained and extended what is now 23 straight years of patent leadership. Most notably, Dr. Kelly and his team were responsible for advancing the science of cognitive computing through his support for Watson, the groundbreaking system that defeated two standing Jeopardy world champions in 2011. He also led the strategy, negotiations, and transition of IBM's semiconductor manufacturing business to GlobalFoundries, while maintaining a commitment to leading-edge microelectronics research. Prior to joining IBM Research in July 2007, Dr. Kelly was senior vice president of Technology and Intellectual Property, responsible for IBM's technical and innovation strategies.

KATHLEEN R. McKEOWN is the director of the Data Science Institute and the Henry and Gertrude Rothschild Professor of Computer Science at Columbia University. She served as department chair from 1998-2003 and as Vice Dean for Research for the School of Engineering and Applied Science for 2 years. A leading scholar and researcher in the field of natural language processing, McKeown focuses her research on big data; her interests include text summarization, question answering, natural language generation, and multilingual applications. She has received numerous honors and awards, including AAAI Fellow, a Founding Fellow of the Association for Computational Linguistics, and an ACM Fellow. Early on she received the National Science Foundation Presidential Young Investigator Award and a National Science Foundation Faculty Award for Women. In 2010, she won both the Columbia Great Teacher Award—an honor bestowed by the students—and the Anita Borg Woman of Vision Award for Innovation.

JOSEPH L. MUNDY is president and chief executive officer of Vision Systems, Inc. Dr. Mundy joined General Electric's Research and Development Center (CRD) in 1963. His early projects at CRD include high-power microwave tube design, superconductive computer memory devices, the design of high density integrated circuit associative memory arrays, and the application of transform coding to image data compression. He is the co-inventor of varactor bootstrapping, a key technique still widely used today in the design of CMOS integrated circuits. From 1972 until 2002, Dr. Mundy led a group involved in the research and development of image understanding and computer vision systems. In the early 1970s his group developed one of the first major applications of computer vision to industrial inspection. A system was developed to inspect incandescent lamp filaments at the rate of 15 parts/second and achieved classification performance of less than one error per thousand. The system operated in production for many years. His more recent research themes at CRD included industrial photogrammetry for machine control, theory of geometric invariance, change detection in satellite imagery, and CT classification of lung cancer lesions. In 1988 Dr. Mundy was named a Coolidge Fellow, GE's highest technical honor. He applied the fellowship to a sabbatical at Oxford University, working with Sir Michael Brady and Professor Andrew Zisserman on applications of invariant theory to computer vision. This work led to the Marr Prize award in 1993. In 2002 Dr. Mundy joined the School of Engineering at Brown University as professor engineering (research). His research at Brown, under sponsorship of DARPA and NGA, includes video and image analysis, with emphasis on change detection and 3D volumetric modeling. In 2011 Dr. Mundy co-founded Vision Systems, Inc., and is president and chief executive officer. At VSI, Dr. Mundy has managed the DARPA Tailwind and Visual Media Reasoning projects that are aimed at aerial video processing and scene analysis. Dr. Mundy also provides general project management of current VSI efforts in 3D modeling from satellite imagery, geo-location of ground-level imagery, and facial recognition. Dr. Mundy received his B.E.E. (1963), M.Eng. (1966), and Ph.D. (1969) from Rensselaer Polytechnic Institute. He has published over 100 papers and articles in computer vision and solid-state electronics.

B

Workshop Agenda

AUGUST 9-10, 2017

The Keck Center of the National Academies of
Sciences, Engineering, and Medicine
Washington, D.C.

Objectives

Research Challenges:
1. Machine-based methods for generating analytic products.
2. Machine-based methods for automating the evaluation of analytic products.

Research Questions:
1. What are the technical objectives and metrics needed for success?
2. What are the primary issues?
3. What are the current and "next level" key performance metrics?
4. What is the "level after next" of expected research and development performance?
5. What is the research knowledge base?
6. How can the government best prepare the scientific workforce to enhance discovery in this area?
7. What are the requisite enabling technologies?

Day 1: August 9, 2017

7:30 A.M. Registration and breakfast (on your own)

SESSION 1: Plenary

8:00 Sponsor Remarks and Expectations of the Workshop
Dr. David M. Isaacson, ODNI

8:15 Generation of Capability Technology Matrix
Dr. Rama Chellappa, UMCP, Planning Committee Chair
Dr. George Coyle, RSO, AFSB/ICSB

8:30	Progress in Machine Learning Dr. Tom Dietterich, Oregon State University
9:05	Industry Perspective Dr. Josyula R. Rao, Watson IBM Fellow
9:45	Operational Perspective—Project MAVEN Dr. Travis W. Axtell, OSD OUSD (I)
10:25	Break

SESSION 2: Machine Learning from Image/Video/Map Data

10:45	Learning from Overhead Imagery Dr. Joe Mundy, Vision Systems, Inc.
11:20	Deep Learning for Learning from Images and Videos: Is It Real? Dr. Rama Chellappa, UMCP
11:55	Learning about Human Activities from Images and Videos Dr. Anthony Hoogs, Kitware, Inc.
12:30 P.M.	Lunch

SESSION 3: Machine Learning from Natural Languages (ML-NLP)

1:15	Machine Learning from Text: Applications Dr. Kathy McKeown, Columbia University
1:50	Deep Learning for NLP Dr. Dragomir Radev, Yale University
2:25	Machine Learning from Conversational Speech Dr. Amanda Stent, Bloomberg
3:00	Break

SESSION 4: Learning from Multi-Source Data

3:15	Situational Awareness from Multiple Unstructured Sources Dr. Boyan Onyshkevych, DARPA
3:50	Discussion on Preparing the Capability Matrix Compile enabling technologies from 1st Day
5:30	Adjourn

Day 2: August 10, 2017

7:30 A.M.	Breakfast in cafeteria (on your own)
8:00	Sponsor Remarks Dr. David Honey, Director of Science & Technology, ODNI

SESSION 5: Learning from Noisy, Adversarial Inputs

8:15	Harnessing Machine Learning for Global Discovery at Scale Dr. Mikel Rodriguez, MITRE

SESSION 6: Learning from Social Media

8:50	Large-Scale Multi-Modal Deep Learning Dr. Rob Fergus, NYU
9:25	What Can We Learn from Social Media Posts? (Presentation withdrawn)
10:00	Break

SESSION 7: Humans and Machines Working Together with Big Data

10:15	Sensemaking Systems and Models Dr. Peter Pirolli, Institute for Human and Machine Cognition
10:50	Crowd Sourcing for Natural Language Processing Dr. Chris Callison-Burch, University of Pennsylvania

SESSION 8: Use of Machine Learning for Privacy Ethics

11:25	Toward Socio-Cultural Machine Learning Dr. Mark Riedl, Georgia Institute of Technology
12:00 P.M.	Lunch

SESSION 9: Panel on Evaluation of Machine-Generated Products

1:00	Dr. Anthony Hoogs, Kitware Dr. Jason Duncan, MITRE Mr. Jonathan Fiscus, NIST Dr. Rob Fergus, NYU
2:00	Break

SESSION 10: Capability Technology Matrix Panel: NSF, DoD, NIST, DoE

2:10	Machine Learning for Energy Applications Dr. Devanand Shenoy, DOE
2:30	Using Metrology to Improve Access to "Unstructured" Data Dr. Ellen Voorhees, NIST
2:50	Challenge Problems for Multi-Source Insights Dr. Travis W. Axtell, OSD OUSD (I)
3:10	An Overview of NSF Research in Data Analytics Mr. James Donlon, NSF
3:30	Discussion on Preparing the Capability Matrix Compile enabling technologies from 2nd day Complete matrix
5:00	Adjourn

C

Workshop Statement of Task

The objective of this effort is to provide insights from world-class experts and technologists familiar with the extensive range of issues associated with the application of artificial intelligence to the Intelligence Community (IC) analytic tradecraft. Existing and emerging capabilities, technical implications, cultural, regulatory, policy, and legal issues are all of interest. This effort will require arranging and running a two-day workshop involving a wide range of technical disciplines in industry, academia, and government to investigate issues around the application of artificial intelligence techniques to IC analytic tradecraft. This unclassified workshop will address the following research challenges: (1) Machine-based methods for generating analytic products, and (2) Machine-based methods for automating the evaluation of analytic products. The workshop will explore the state-of-the-art and security implications of existing and emerging technologies pertinent to these topics. A workshop proceedings will be delivered after the workshop. The following tasks are anticipated:

Task 1: Form an ad hoc planning committee of experts to be approved by the leadership of the National Academies. The committee will include world-class experts and technologists familiar with the broad range of issues germane to the application of artificial intelligence techniques to IC analytic tradecraft. When developing the agenda for the workshop, the committee will consider the following research questions:

- What are the technical objectives and metrics needed for success?
- What are the primary issues?
- What are the current and "next level" key performance metrics? What is the "level after next" of expected research and development performance?
- What is the research knowledge base?
- How can the government best prepare the scientific workforce to enhance discovery in this area?

Task 2: Arrange and execute a workshop of at least two days duration. This includes securing a venue, organizing and managing the workshop, and providing adequate facilitators and note takers to ensure that prepared materials and insights from the workshop are saved.

Task 3: Prepare a workshop proceedings, including presenters' materials, that records the experts' key arguments, information, and insights presented during the course of the workshop . The final reviewed draft proceedings will undergo sponsor security review before posting or otherwise publishing in whole or in part.

D

Capability Technology Tables

TABLE D.1 Current Capabilities

Presenter	Current Capabilities
Tom Dietterich, Oregon State University	• Apply deep learning to signals-type data if they are stationary and if there is enough training data • Adapt quickly to new problems using fine tuning and, to some extent, anomaly detection • Participate in rapidly growing Kaggle challenges across a huge range of problem domains
Joseph Mundy, Vision Systems, Inc.	• Global-scale object detection and recognition in satellite image enabling broad-area search anywhere on the Earth's surface • End-to-end deep net production of three-dimensional stereo reconstructions yielding significant improvement in disparity accuracy and completeness • Application of the AlphaGo game search strategy to other rule-driven processes such as geometric modeling, a promising avenue to inject theory into data-driven methods
Rama Chellappa, University of Maryland, College Park	• Object detection with much improved performance for thousands of objects • Face detection with much improved performance for frontal faces and faces with >15 pixels • Object recognition with much improved performance for well-defined objects • Face recognition/verification with 91 percent TAR at 10-4 FAR for face verification on JANUS–CS4 data set—Can handle 1 million-sized gallery
Dragomir Radev, Yale University	• Integrating deep learning and reinforcement learning (cf. AlphaGo) • Automating architecture learning • Hyperparameter search • Improving generative models • Memory-based networks and neural Turing machines • Semi-supervised learning • Mix-and-match architectures • Privacy-preserving models

continued

TABLE D.1 Continued

Presenter	Current Capabilities
Peter Pirolli, Institute for Human and Machine Cognition	• Commercial off-the-shelf interactive visual analytics now pervasive for big data analysis • Mature research platforms for visualization grammars and toolkits • Early versions of interactive visualization for machine learning experts and learners (e.g., TensorFlow, Playground) • Emerging Standard Model of Cognition for constrained, reasonably well-defined tasks/domains (limited transfer/generalization) • Cognitive models of users/students in long-term tutoring with capabilities for induction/fine-tuning of student models • User modeling and personalization is a mature field for well-defined tasks/domains • Models of joint interdependent activity in non-learning tasks (e.g., human-robot interaction) • Interactive task learning for constrained robot tasks and contexts • Programming by demonstration • Mobile phone digital assistants
Chris Callison-Burch, University of Pennsylvania	• Recent emergence of start-ups focused on annotation such as CrowdFlower, which is being leveraged by the Department of Defense and the intelligence community
Mark Riedl, Georgia Institute of Technology	• Value alignment is an unsolved problem—no examples of aligned agents except in toy worlds • Virtually no understanding of sociocultural conventions that reduce human–human conflict • Can learn commonsense knowledge on curated corpora in limited a priori known contexts • Currently data is bottleneck employing curated corpora—deep learning is needed to learn human values/conventions from stories on the Internet • Curated data sets do not currently exist, and researchers still do not know how to solve these problems using deep neural networks
Panel on Evaluation of Machine-Generated Products; Anthony Hoogs, Panel Moderator	• Robust activity from Intelligence Advanced Research Projects Activity in creating data challenges • Widespread public awareness of and participation in data notation and labeling (e.g., Mechanical Turk)

TABLE D.2 Short-Term (3–5 years) Capabilities

Presenter	Short-Term (3–5 years) Capabilities
Tom Dietterich, Oregon State University	• Open category classification • Automatic detection of biased and untrusted sources • Anomaly detection on time-varying and network data • Initial methods for validation and system monitoring
Joseph Mundy, Vision Systems, Inc.	• The rapid growth of web-linked data sources based on semantic ontologies such as geographic and functional knowledge should be exploited to label training data and to provide context for test time • A major effort in incorporating theoretical knowledge into data-driven decision making • Using machine learning to drive logical and grammatical structure formation is a promising first step • Extend current deep learning three-dimensional modeling algorithms to four dimensions and combine with object recognition networks to achieve functional designs such as building information models
Rama Chellappa, University of Maryland, College Park	• Handle pose, illumination, and low-resolution challenges in unconstrained scenarios • Handle full motion videos • Sharing/transfer learning of features for different but somewhat related tasks • Model/understand context and occlusion • Handle "reasonable" amount of noise in data • Limited robustness to adversarial data • Vision and language • Other multi-modal data
Anthony Hoogs, Kitware, Inc.	• Generative adversarial networks effectively applied to video • Increased model transfer between video domains • Human-level accuracy for action recognition in single-action, temporal clipped videos — UCF 101 — Human-level accuracy achieved for ImageNet object recognition in primary subject images 1,000 categories • Major improvements in action and complex activity recognition in surveillance video — Person tracking and re-identification — Defense Intelligence Agency's IARPA project • Super-human performance in video search and retrieval for objects and primary subjects in internet (consumer) videos — Cataloguing of large, common objects and events — Similar accuracy to humans but much faster — Free-form text-based queries with limited syntax and vocabulary • Large graphical processing unit farms required to keep up with video generation
Dragomir Radev, Yale University	• Language data will be more and more in speech form • Massive multilingulairty • One-shot learning (e.g., Tigrinya and Oromo) • Understanding short texts • Interpretability of models • Deep learning personalization • Trust analytics • Ethical language processing
Mikel Rodriguez, MITRE	• More work needs to be done on short- and long-term prediction; recent models such as Long Short-Term Memory Networks (LSTMs) and recurrent neural networks have just begun to address this gap • More work needs to be done on complex activities, temporal sequences, rare targets, and modalities unique to the Department of Defense/intelligence community, as well as counter artificial intelligence assurance

continued

TABLE D.2 Continued

Presenter	Short-Term (3–5 years) Capabilities
Peter Pirolli, Institute for Human and Machine Cognition	• Multi-level models across cognitive, rational, and social bands of interactive phenomena for well-defined, "stationary" sensemaking tasks • Sensemaking tasks (but still intelligence community-relevant) • Explainable artificial intelligence for visual analytics and simulated drone operations • Interactive task learning for usable soft bots for well-defined tasks (clear goal, well-defined constraints, clear operators) • Improved visual analytics for machine learning programming
Chris Callison-Burch, University of Pennsylvania	• Tighter integration of crowdsourcing with machine learning — Correct/confirm output from models — Active learning — Domain adaptation • New crowdsourcing platform for natural language processing — Remove hassles of Mechanical Turk — Cultivate groups of language experts — Create standing pools of language workers — Deployable on inside of the intelligence community
Mark Riedl, Georgia Institute of Technology	• Agents expected to start using commonsense knowledge and world knowledge to address human needs • Conversational agents + expectation; need to engage in longer conversations and rely on computational imagination • Agents expected to differentiate behavior based on cultural context • Computational creativity as part of mainstream content creation
Panel on Evaluation of Machine-Generated Products; Anthony Hoogs, Moderator	• Open source tools for efficient annotation • Semi-automated labeling to efficiently fuse computed labels with manual adjudication • Large-scale annotation of operationally representative data sets in domains of interest, made available to researchers, particularly multi-modal data sets

NOTE: IARPA, Intelligence Advanced Research Projects Activity.

TABLE D.3 Long-Term (> 5 years) Capabilities

Presenter	Long-Term (>5 years) Capabilities
Tom Dietterich, Oregon State University	• Defense against adversarial examples • Integration of large knowledge bases with machine learning • Use of meta reasoning to develop and test hypotheses about data source reliability • Multi-scale machine learning
Joseph Mundy, Vision Systems, Inc.	• Drive towards global continuous learning from the emerging constellations of overhead data, including aerial video, incorporating adaptive formation of new relations and concepts not present in the original training data. The resulting ontological knowledge will support all forms of intelligence analysis.
Rama Chellappa, University of Maryland, College Park	• Theoretical investigation of why and when deep learning works • Minimum training set needed given empirical distributions of training set in the absence/presence of domain shift; error bounds; incorporating invariances • Image formation, blur, geometry, shape, motion, texture, and occlusion models that can generalize using data • Develop machine learning methods that generalize from small data — Humans generalize well from small data • Develop machine learning methods that can deal with parts, interconnections of parts, adversarial data • Incorporate domain knowledge, common sense reasoning
Anthony Hoogs, Kitware, Inc.	• Structure learning of deep networks on a large scale • Human-level accuracy for action and complex activity recognition in surveillance video — DVA data • Super-human performance in video search and retrieval for complex activities and abstractions in internet (consumer) videos — Cataloguing of all objects, scene elements, and events — Similar accuracy to humans but much faster — Free-form text-based queries with open syntax and vocabulary • Large graphical processing unit farms required to keep up with video generation — Computation speed increases offset by data growth
Kathy McKeown, Columbia University	• Interpretation, analysis, and generation from informal text (social media, online discussion, online narrative), multi-modal sources, and streaming data • Learning without big data (low resource) • Understanding machine learning: explanation/linguistics • Discourse, context • Applications in multi-lingual environments without machine translation • Robust and flexible language generation • Fake news, cyber bullying, inappropriate content

continued

TABLE D.3 Continued

Presenter	Long-Term (>5 years) Capabilities
Amanda Stent, Bloomberg	• Explore and understand: sponsor research on machine learning for conversational analytics — Identifying "problematic" dialogues and when dialogues go wrong • Produce: challenge the community to move beyond (single-domain) slot filling and (context-free) chat — Situated dialogue — Negotiation dialogue — Complex and flexible tasks — Ongoing/perpetual dialogue work at Carnegie Mellon University • Encourage methods that combine advanced machine learning with programmatic control or otherwise infuse machine learning with human knowledge • Encourage methods that include "human-in-the-loop" — Active learning — Just-in-time human computation • Provide the community with shared portals for evaluation (and crowdsourcing of data) • Encourage the development machine learning as a service rather than a puzzle or an art — Where should machine learning live in relation to data and services? — How can machine learning be reproducible? • Sponsor work on world knowledge induction and situational awareness • Collaborate on ameliorating data availability issues — Privacy-preserving methods for mining
Peter Pirolli, Institute for Human and Machine Cognition	• Foundational science of Human-Autonomy Collaboration — Transition from "programming" to "learning to work together" — Sensemaking, autonomous vehicles, precision behavioral medicine, workforce — Education, pervasive personalized assistants — Multi-level models of sensemaking that include dynamically adapting humans and machine learning in joint activity • Continuous in-the-loop explainable artificial intelligence and interactive task learning — Usable, scalable, generalizable, dynamic interactive task learning • Evaluation frameworks on joint human–autonomy tasks — Move beyond current machine learning-centric optimization metrics • Open source data and code for tasks that have some relevance and validity to the intelligence community
Mark Riedl, Georgia Institute of Technology	• Increasing presence of autonomous systems in social contexts — Chicken-and-egg problem; the problem of humans understanding artificial intelligence and artificial intelligence understanding humans will be far from solved — Explainable artificial intelligence • Broader artificial intelligence safety concerns and "big red button problem" • Use of military teammates could become more common • Computational creativity: Autonomous creation of images and video that are difficult to detect

E

Acronyms

AIDA	Active Interpretation of Disparate Alternatives
CISE	Directorate for Computer and Information Science and Engineering
DARPA	Defense Advanced Research Projects Agency
DEFT	Deep Exploration and Filtering of Text
DIVA	Deep Intermodal Video Analytics
DOD	Department of Defense
DOE	Department of Energy
IARPA	Intelligence Advanced Research Projects Activity
ICSB	Intelligence Community Studies Board
LORELEI	Low Resource Languages for Emergent Incidents
MediFor	Media Forensics
NIST	National Institute of Standards and Technology
NSF	National Science Foundation
ODNI	Office of the Director of National Intelligence
OUSD(I)	Office of the Under Secretary of Defense for Intelligence
TREC	Text REtrieval Conference